# ENGINEERING PROJECT MANAGEMENT

Second Edition

Edited by

**Nigel J. Smith**
*Professor of Construction Project Management*
*University of Leeds*

**Blackwell**
Science

© 2002 by Blackwell Science Ltd,
a Blackwell Publishing Company
Editorial Offices:
Osney Mead, Oxford OX2 0EL, UK
  Tel: +44 (0)1865 206206
Blackwell Science, Inc., 350 Main Street,
Malden, MA 02148-5018, USA
  Tel: +1 781 388 8250
Iowa State Press, a Blackwell Publishing Company,
2121 State Avenue, Ames, Iowa 50014-8300, USA
  Tel: +1 515 292 0140
Blackwell Publishing Asia Pty Ltd, 550 Swanston
Street, Carlton South, Melbourne, Victoria 3053,
Australia
  Tel: +61 (0)3 9347 0300
Blackwell Wissenschafts Verlag, Kurfürstendamm 57,
10707 Berlin, Germany
  Tel: +49 (0)30 32 79 060

First edition published 1995 by Blackwell Science Ltd
Reprinted 1996, 1998

Second edition published 2002 by Blackwell Science
Ltd

Library of Congress
Cataloging-in-Publication Data
Engineering project management/edited by Nigel J.
  Smith.—2nd ed.
    p. cm.
  Includes bibliographical references and index.
  ISBN 0-632-05737-8 (alk. paper)
    1. Engineering—Management. 2. Construction
  industry—Management. I. Smith, Nigel J.

TA190 .E547 2002
658.4′04—dc21

                                          2002074568

ISBN-0-632-05737-8

A catalogue record for this title is available from the
British Library

Set in 10/13pt Times
by DP Photosetting, Aylesbury, Bucks
Printed and bound in Great Britain by
TJ International, Padstow, Cornwall

For further information on
Blackwell Science, visit our website
www.blackwell-science.com

# Contents

# Preface

In many sectors of industry the significance of good project management in delivering projects in accordance with predetermined objectives has been established. Industrialists and engineering institutions have called for the inclusion of a significant proportion of project management in higher-level degrees, something realised by Finniston in his review of the future of engineering in 1980 (Engineering our Future. Report of the Committee of Inquiry into the Engineering Profession, Chairman Sir Montague Finniston, HMSO, 1980). Since the publication of the first edition of this book in 1995, a number of significant developments have taken place. A British Standard for Project Management, BS6079, has been published, and the UK-based Association for Project Management has produced a fundamental guide to processes and practice entitled *Body of Knowledge*, and has drafted a standard contract for employing project managers. There has also been a marked increase in the teaching and delivery of university programmes and continuing professional development (CPD) courses in project management.

Many organisations in the engineering, finance, business, process and other sectors are appointing project managers. Some have a very narrow brief and a precise role, whereas others have a more strategic, managerial and multi-disciplinary function. This second edition builds upon the success of the first edition in providing a clear picture of the aims of project management based upon best practice. The improvements in this edition have been driven by changes in the practice of project management, and by the helpful comments made by book reviewers and readers since 1995.

The original information on risk management is updated and enhanced. The principle of uncertainty management is recognised by a new chapter on value management to balance the effective management of adverse risk and opportunity. Changes in the management of major projects have resulted in more joint ventures, project partnering, special project vehicles and other forms of collaborative working. The new edition includes new chapters on supply and value chain management, and on effective project partnering. The book is not aimed at any particular sector of engineering, and relates to the management of any major technical project.

Newly appointed project managers and students of project management at the MEng, MBA and MSc level will find the enhanced text and references beneficial. The book is concerned with the practice and theory of project management, particularly in relation to multi-disciplinary engineering projects, large and small, in the UK and overseas.

Nigel J. Smith

# List of Contributors

**Editor: Professor Nigel J. Smith**, BSc, MSc, CEng, FICE, MAPM, is Professor of Construction Project Management in the School of Civil Engineering, University of Leeds. After graduating from the University of Birmingham he has spent fifteen years in the industry working mainly on transportation infrastructure projects. His academic research interests include risk management and procurement of projects using private finance. He has published fifteen books and numerous refereed papers. He is currently Dean of the Faculty of Engineering.

**Denise Bower**, BEng, PhD, MASCE, is a Lecturer in Project Management in the School of Civil Engineering at the University of Leeds. She was formerly the Shell Lecturer in Project Management at UMIST. She is a leading member of Engineering Management Partnership which offers Diploma and Masters Degree level qualifications to engineers of all disciplines. She was a member of Latham Working Group 12 and has an extensive record of consultancy work with clients in construction, process engineering and manufacturing.

**Peter Harpum**, after ten years managing major projects in the oil, shipping and construction industries, Peter Harpum now works as a project management consultant advising blue chip clients in various industrial sectors. He has carried out a number of research projects with international engineering and construction organisations, concentrating in particular on the front-end of the project lifecycle. He has contributed to the design, content, and delivery of a number of project management courses at Masters level and continues to work as a visiting lecturer at UMIST. Peter holds a Masters Degree in Project Management, and is working on his doctoral thesis on design management theory.

**Professor Steven Male**, holds the Balfour Beatty Chair in Building Engineering and Construction Management, School of Civil Engineering, University of Leeds. His research and teaching interests include strategic management in construction, supply chain management, value management and value engineering. He has led research projects under the EPSRC IMI programme 'Construction as a Manufacturing Process'; with the DETR; DTI; and, within the European Union 4th and 5th Frameworks. He is a visiting Professor in the Department of Civil Engineering, University of Chile. He works closely with industry and has undertaken a range of research, training and consultancy studies with construction corporations, construction consultancy firms, blue chip and government clients.

**Tony Merna**, BA, MPhil, PhD, CEng, MICE, MAPM, MIQA, is the senior partner of the Oriel Group Practice and is also a Lecturer in the Department of Civil and Construction Engineering at UMIST. A chartered engineer by profession, he has been involved at the design, construction, operation and financing stages of many projects procured in the UK and overseas.

**Kris Moodley**, BSc, MSc, AIArb, is a Lecturer in Construction Management in the School of Civil Engineering, University of Leeds. After graduating from Natal University, his initial employment was with Farrow Laing in Southern Africa, before his first academic appointment at Heriot Watt University. He moved to Leeds four years later to pursue research interests in strategic business relationships between organisations and their projects. He has contributed to many publications and is co-author of the book *Corporate Communications in Construction*.

**Ian Vickridge**, BSc(Eng), MSc, CEng, MICE, MCIWEM, runs his own civil engineering consultancy and is also a Visiting Senior Lecturer in Civil Engineering at UMIST. He has over thirty years experience in the construction industry gained in Canada, Hong Kong, Singapore, China, and Saudi Arabia as well as the UK. He is the Executive Secretary of the UK Society for Trenchless Technology and a reviewer of candidates for Membership of the Institution of Civil Engineers. He has published extensively on a variety of topics related to construction, environmental management and project management.

**David Wright**, MA, CIChemE, ACIArb, left Oxford with a degree in Jurisprudence and spent 30 years in industry. He gained experience in the automotive industry, the electronic industry, the defence industry and the chemical engineering and process industry. He was Commercial Manager of Polibur Engineering Ltd and in the mechanical engineering sector was European Legal Manager to the Mather & Platt Group. He is now a consultant on matters of contract and commercial law. David is also a Visiting Lecturer at UMIST and a Visiting Fellow in European Business Law at Cranfield University.

# Acknowledgements

I am particularly grateful to my co-authors and fellow contributors to this book, especially those who have participated in both editions, namely Dr Denise Bower, Dr Tony Merna and Mr Ian Vickridge. I am also grateful to the new contributors Mr Pete Harpum, Professor Steve Male, Mr Krisen Moodley and Mr David Wright. I would also like to thank again the contributors to the first edition, particularly Professors Peter Thompson and Steven Wearne; also Dr Kareem Yusuf of IBM for his new contribution in software and modelling in Chapter 9.

The editor and authors would like to express their appreciation to Sally Mortimer for managing the existing artwork from the first edition, and new diagrams from the authors supplied in a wide variety of formats for processing. I would also like to thank Sally for checking and revising each of the many draft versions of every chapter. Nevertheless, the responsibility for any errors remains entirely my own.

Nigel J. Smith

# List of Abbreviations

| | |
|---|---|
| ABS | Assembly breakdown structure |
| ACWP | Actual cost of work performed |
| ADB | Asian Development Bank |
| ADR | Alternative dispute resolution |
| AfDB | African Development Bank |
| APM | Association for Project Management |
| BAC | Budget (baseline) at completion |
| BCWP | Budgeted cost of work performed |
| BCWS | Budgeted cost of work scheduled |
| BOD | Build–operate–deliver |
| BOL | Build–operate–lease |
| BOO | Build–own–operate |
| BOOST | Build–own–operate–subsidise–transfer |
| BOOT | Build–own–operate–transfer |
| BoQ | Bills of quantities |
| BOT | Build–operate–transfer |
| BPR | Business process re-engineering |
| BRT | Build–rent–transfer |
| BTO | Build–transfer–operate |
| CBA | Cost–benefit analysis |
| CII | Construction Industry Insitute (Texas) |
| CPD | Continuing professional development |
| CPI | Cost performance index |
| CV | Cost variance |
| DBOM | Design–build–operate–maintain |
| DBOT | Design–build–operate–transfer |
| DCMF | Design, construct, manage and finance |
| DEO | Defence Estates Organisation |
| DETR | Department of the Environment, Transport and the Regions |
| DFA | Design for assembly |
| DfID | Department for International Development |

| DFM | Design for manufacturing |
| DSM | Dependency structure matrix |
| DTI | Department of Trade and Industry |
| EBRD | European Bank for Reconstruction and Development |
| ECC | Engineering and construction contract |
| ECGD | Export Credit Guarantee Department |
| ECI | European Construction Institute |
| EIA | Environmental impact assessment |
| EIB | European Investment Bank |
| EIS | Environmental impact statement |
| EMS | Environmental management system |
| EPC | Engineer–procure–construct |
| EPIC | Engineer–procure–install–commission |
| EQI | Environmental quality index |
| ERP | Enterprise resource planning |
| EU | European Union |
| EVA | Earned value analysis |
| FBOOT | Finance–build–own–operate–transfer |
| FIDIC | Fédération Internationale des Ingénieurs Conseils (Lausanne) |
| GUI | Graphical user interface |
| HMPS | Her Majesty's Prison Service |
| IFC | International Finance Corporation |
| IPT | Integrated project team |
| IRR | Interest rate risk |
| MoD | Ministry of Defence |
| NEPA | National Environmental Protection Agency |
| NGO | Non-governmental organisation |
| NIF | Note issuance facility |
| NPV | Net present value |
| OBS | Organisational breakdown structure |
| OECD | Organisation for Economic Cooperation and Development |
| PBS | Product breakdown structure |
| PC | Procure–construct |
| PEP | Project execution plan |
| PERT | Programme Evaluation and Review Technique |
| PFI | Private finance initiative |
| PIC | Procure–install–commission |
| PIM | Personal information manager |
| PMI | Project Management Institute |
| PSBR | Public sector borrowing requirement |
| QA | Quality assurance |

| QC | Quality control |
|------|-----------------|
| QFD | Quality function deployment |
| RUF | Revolving underwriting facility |
| SCA | Structured concession agreement |
| SCM | Supply chain management |
| SPI | Schedule performance index |
| SPV | Special project vehicle |
| SV | Schedule variance |
| TCM | Travel cost method |
| TCN | Third country nationals |
| TQM | Total quality management |
| TUPE | Transfer of undertaking from previous employer |
| USGF | US Gulf factor |
| VA | Value analysis |
| VE | Value engineering |
| VM | Value management |
| VP | Value planning |
| VR | Value reviewing |
| WBS | Work breakdown structure |
| WMG | Warwick Manufacturing Group |
| WTA | Willingness to accept |
| WTP | Willingness to pay |

# Chapter 1
# Projects and Project Management

This chapter describes the various aspects of project management from what a project is, through its various stages, to the key requirements for success.

## 1.1 The function of project management

Managing projects is one of the oldest and most respected accomplishments of mankind. One stands in awe of the achievements of the builders of the pyramids, the architects of ancient cities, the masons and craftsmen of great cathedrals and mosques, and of the might of labour behind the Great Wall of China and other wonders of the world. Today's projects also command attention. People were riveted at the sight of the Americans landing on the moon. As a new road or bridge is opened, as a major building rises, as a new computer system comes online, or as a spectacular entertainment unfolds, a new generation of observers is inspired.

All of these endeavours are projects, like many thousands of similar task-orientated activities, yet the skills employed in managing projects, whether major ones such as those mentioned above or more commonplace ones, are not well known other than to the specialists concerned. The contribution that a knowledge of managing projects can make to management at large is greatly underrated and generally poorly known. For years, project management was derided as a low-tech, low-value, questionable activity. Only recently has it been recognised as a central management discipline. Major industrial companies now use project management as their principal management style. 'Management by projects' has become a powerful way to integrate organisational functions and motivate groups to achieve higher levels of performance and productivity.

## 1.2 Projects

A project can be any new structure, plant, process, system or software, large or small, or the replacement, refurbishing, renewal or removal of an existing one. It is a one-off investment. In recent times, project managers have had to meet the demands of increasing complexity in terms of technical challenge, product sophistication and organisational change.

One project may be much the same as a previous one, and differ from it only in detail to suit a change in market or a new site. The differences may extend to some novelty in the product, in the system of production, or in the equipment and structures forming a system. Every new design of car, aircraft, ship, refrigerator, computer, crane, steel mill, refinery, production line, sewer, road, bridge, dock, dam, power station, control system, building or software package is a project. So are many smaller examples, and a package of work for any such project can in turn be a subsidiary project.

Projects thus vary in scale and complexity from small improvements to existing products to large capital investments. The common use of the word 'project' for all of these is logical because every one is:

❏   an investment of resources for an objective;
❏   a cause of irreversible change;
❏   novel to some degree;
❏   concerned with the future;
❏   related to an expected result.

A project is an investment of resources to produce goods or services; it costs money. The normal criterion for investing in a proposed project is therefore that the goods or services produced are more valuable than the predicted cost of the project.

To get value from the investment, a project usually has a defined date for completion. As a result, the work for a project is a period of intense engineering and other activities, but is short in its duration relative to the subsequent working life of the investment.

A number of definitions of the term 'project' have been proposed, and some are presented below.

❏   The Project Management Institute (PMI), USA, defines a project as 'a temporary endeavour undertaken to create a unique product or service'.
❏   The UK Association for Project Management defines a project as 'a discrete undertaking with defined objectives often including time, cost and quality (performance) goals'.

❑ The British Standards Institute (BS6079) defines a project as 'a unique set of coordinated activities, with definite starting and finishing points, undertaken by an individual or organisation to meet specific objectives with defined schedule, cost and performance parameters'.

From the above definitions, it may be concluded that a project has the following characteristics:

❑ it is temporary, having a start and a finish;
❑ it is unique in some way;
❑ it has specific objectives;
❑ it is the cause and means of change;
❑ it involes risk and uncertainty;
❑ it involves the commitment of human, material and financial resources.

## 1.3 Project management

The definition of project management stems from the definition of a project, and implies some form of control over the planned process of explicit change.

❑ The PMI defines project management as 'the art of directing and coordinating human and material resources through the life of a project by using modern management techniques to achieve predetermined goals of scope, cost, time, quality and participant satisfaction'.
❑ The UK Association for Project Management defines it as 'the planning, organisation, monitoring and control of all aspects of a project and the motivation of all involved to achieve project objectives safely and within agreed time, cost and performance criteria'.
❑ The British Standards Institute (BS 6079) defines it as 'the planning, monitoring and control of all aspects of a project and the motivation of all those involved to achieve the project objectives on time and to cost, quality and performance'.

The common theme is that project management is the management of change, but explicitly planned change, such that from the initial concept, the change is directed towards the unique creation of a functioning system. In contrast, general or operations management also involves the management of change, but their purpose is to minimise and control the effects of change in an already constructed system. Therefore, project management directs all the elements that are necessary to reach the objective, as well as those that will hinder the development. It should not be forgotten that projects are managed with and through people.

Project management must look ahead at the needs and risks, communicate the plans and priorities, anticipate problems, assess progress and trends, get quality and value for money, and change the plans if necessary to achieve the objectives.

The needs of project management are dependent upon the relative size, complexity, urgency, importance and novelty of a project. The needs are also greater where projects are interdependent, particularly those competing for the same resources.

Each project has a beginning and an end, and hence it is said to have a life-cycle. A typical life-cycle is defined by Wearne (1995) and shown in Figure 1.1. The nature and scale of activity change at each stage. Each stage marks a change in the nature, complexity and speed of the activities and the resources employed as a project proceeds. Widely different terminology for the various aspects of project management are used in different industries, but they can all be related to this diagram.

The durations of the stages vary from project to project, and delays sometimes occur between one stage and the next. Stages can also overlap. Figure 1.1 shows the sequence in which the stages should be started. This is not meant to show that one must be completed before the next is

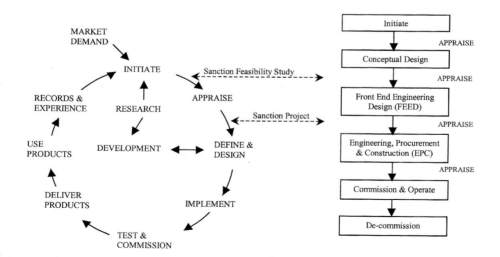

**Figure 1.1** Project life cycle.

started, however. The objective of the sequence should be to produce a useful result, so that the purpose of each stage should be to allow the next to proceed.

## 1.4 Project initiation

As shown in Figure 1.1, a project is likely to be initiated when its promoter predicts that there will be a demand for the goods or services the project will produce. The planners of the project should then draw on the records and experience of previous projects, and the results from research indicating new possibilities. These three sources of information ought to be brought together at this stage of a project.

At this stage, there will usually be alternative ideas or schemes which all seem likely to meet the demand. Further progress requires the promoter to authorise the use of some resources to investigate these ideas and the potential demand for the project. The term 'sanction feasibility study' is used here to mean the decision to incur the cost of these investigations. The next move is to appraise the ideas in order to compare their predicted cost with their predicted value.

In emergencies, appraisal is omitted, or if a project is urgent there may be no time to try to optimize the proposal. More commonly, alternative proposals have to be evaluated in order to decide whether to proceed, and then how best to do so in order to achieve the promoter's objectives.

### Feasibility study

The appraisal stage is also known as the 'feasibility study'. The outputs from this phase can only be probabilistic, as they are based upon predictions of demand and costs whose reliability varies according to the quality of the information used, the novelty of the proposals, and the amount and quality of the resources available to investigate any risks that could affect the project and its useful life. The key feature of this phase is the decision as to whether or not the project is viable – that is, high risk levels do not necessarily mean that the project will not go ahead, but rather that a higher rate of return is required.

After the first appraisal, a repetition of the work up to this point is often needed as its results may show that better information is needed on the possible demand, or the conclusions of the appraisal may be disappointing and revised ideas are needed which are more likely to meet the demand. In addition, more expenditure has to be sanctioned. Repetition of the work may also be needed because the information used to predict the demand for the project has changed during the time taken to complete the work. Feasibility studies may therefore have to be repeated several times.

Concluding this work may take time, but its outcome will be quite specific: the sanctioning or rejection of the proposed project. If a project is selected, the activities change from assessing whether it should proceed to deciding how best it should be realised and to specifying *what* needs to be done.

### Design, development and research

Design ideas are usually the start of possible projects, and alternatives are investigated before estimating costs and evaluating whether to proceed any further. The main design stage of deciding how to use materials to realise a project usually follows an evaluation and selection of those materials, as indicated in Figure 1.1. The decisions made at the design stage almost entirely determine the quality and cost, and therefore the success, of a project. Scale and specialisation increase rapidly as the design proceeds.

Development in the cycle involves both experimental and analytical work to test the means of achieving a predicted performance. Research ascertains the properties and potential performance of the materials used. These two are distinct in their objectives. Design and development share one objective, that of making ideas succeed. Their relationship is therefore important, as indicated in Figure 1.1.

Usually, most of the design and the supporting development work for a project follows the decision to proceed. This work may be undertaken in sub-stages in order to investigate novel problems and review the predictions of cost and value before continuing with a greater investment of resources.

### Project implementation

Then follows the largest scale of activities and the variety of physical work needed to implement a project, particularly the manufacture of equipment and the construction work necessary for its completion.

Most companies and public bodies who promote new capital expenditure projects employ contractors and sub-contractors from this stage on to supply equipment or carry out construction. For internal projects within firms there is an equivalent internal process of placing orders to authorise expenditure on labour and materials.

The sections of a project can proceed at different speeds in the design and subsequent stages, but all must come together to test and commission the resulting facility. The project has then reached its productive stage, and should be meeting the specified objectives.

The problems encountered in meeting objectives vary from project to project. They vary in content and in the extent to which experience can be adapted from previous projects in order to avoid novel problems. The criteria for appraisal also vary from industry to industry, but common to all projects is the need to achieve a sequence of decisions and activities, as indicated in Figure 1.1.

Figure 1.1 is a model of what may be typical of the sequence of work for one project. Projects are rarely carried out in isolation from others. At the start, alternative projects may be under consideration, and in the appraisal stage these compete for selection. Those selected are then likely to share design resources with others which may be otherwise unconnected, because of the potential advantages of sharing expertise and other resources, but would therefore be in competition with them for the use of these resources. This sharing will be similar through all the subsequent stages in the cycle.

A project is thus likely to be cross-linked with others at every stage shown in Figure 1.1. These links enable people and firms to specialise in one stage or sub-stage of the work for many projects. The consequence is often that any one project depends upon the work of several departments or firms, each of which is likely to be engaged on a variety of projects for a variety of customers. In all of these organisations there may therefore be conflicts in utilising resources to meet the competing needs of a number of projects, and each promoter investing a project may have problems in achieving the sequence of activities which best suits his interests.

## 1.5  Project risks

Projects are investments of resources, with a distinct increase in the level of investment as the project passes from concept to implementation. This is demonstrated by the 'typical investment curve' shown in Figure 1.2. From this graph, deviations from the base investment profile (i.e. risks) can be identified. Line A indicates increased income, e.g. better than expected sales price/volume, or lower operational costs. Line B indicates completion, e.g. the project is completed late, but the net revenue is then as forecast. Line C indicates reduced net revenue, e.g. worse sales price/volume, or higher operational costs.

All projects are subject to risk and uncertainty. When you purchase

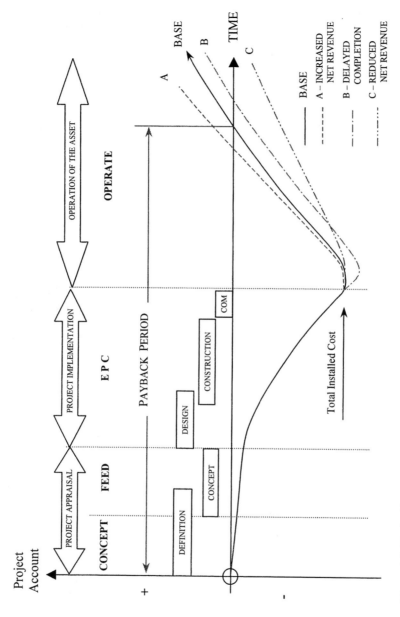

**Figure 1.2** Typical investment curve.

goods from a retailer, you are able to view them before purchase to ensure that they meet your requirements. In other words, you are able to view the finished product prior to making your investment. Unfortunately, this situation is not possible in projects where the promoter is required to make an investment prior to receipt of the product. Accordingly, projects are subject to uncertainty and consequent risk during the project delivery process. Such risks may be generated from factors external to the project (e.g. political change, market demand, etc.), or internally from the activities of the project (e.g. delays due to the weather or unforeseen conditions, etc.). The nature of risk is that it can have both positive and negative effects on the project, in other words, there are said to be upside and downside risks.

## 1.6 Project objectives

It is at the front end of the project cycle that the greatest opportunity exists for influencing the project outcome. This principle is illustrated in Figure 1.3. The curves indicate that it is during the definition and concept stages that the greatest opportunity exists to reduce cost or to add value to the project. This opportunity diminishes as the project passes through sanction to implementation because as more decisions are taken, the project becomes more closely defined. Conversely, the costs of introducing change are magnified, and hence the best time to explore

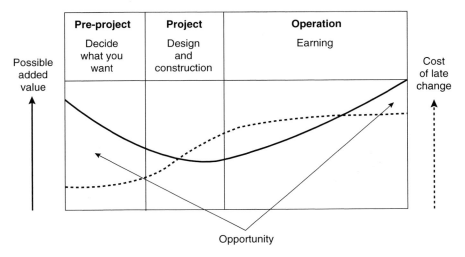

**Figure 1.3** Opportunity to add value.

options and make changes is at the concept stage and certainly not during implementation.

Projects are implemented to meet the objectives of the promoter and the project stakeholders. The term stakeholders is being used here to mean those groups or individuals who have a vested interest in the project, but may or may not be investors in it. Accordingly, it is important that the project's objectives are clearly defined at the outset, and the relative importance of these objectives is clearly established. Primary objectives are usually measured in terms of time, cost and quality, and their inter-relationship is shown in Figure 1.4. The use of an equilateral triangle in this context is significant, since whilst it may be possible to meet one or two of the primary objectives, meeting all three is almost impossible. The positioning of the project in relation to the primary objectives is a matter of preference; where early completion is required, then time is dominant, as might be the case with the launch of a new product where it is necessary to obtain market penetration first. That is not to say that the project can be completed at any cost, and nor does it mean that the quality can be compromised, but time is of the essence. Where minimum cost development is required, such as for a community housing project, quality and time may have to be 'sacrificed'.

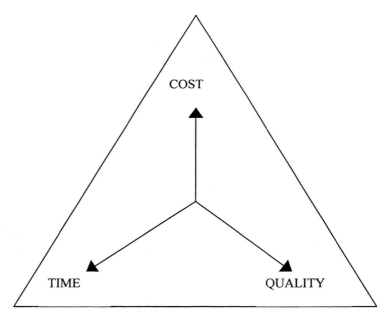

**Figure 1.4** The triangle of project objectives (adapted from Barnes and Wearne, 1993).

Where ultimate quality is required, as, for example, in high-technology projects, then cost and time may be secondary issues.

The relative importance of each objective must be given careful consideration because decisions throughout the project will be based on the balance between them. Inadequate definitions and the poor communication of objectives are common causes of failure in projects. An alignment meeting should be held with all key staff to ensure that all decisions are optimised in terms of the project objective.

## 1.7  Project success

The evidence indicates that the success of projects now and in the future may depend upon the factors listed below.

### Definition of project objectives
The greatest lesson of project management is that the first task is to establish, define and communicate clear objectives for every project.

### Risks
To succeed, the promoter's team should then assess the uncertainties of meeting the project objectives. If the risks are not identified, success cannot be achieved. The volume of events taking the team by surprise will be just too great for them to have any chance of meeting the objectives.

### Early decisions
Many project successes demonstrate the value of completing much of the design and agreeing a project execution plan before commitment to the costly work of manufacturing hardware or constructing anything on-site.

### Project planning
The form and amount of planning have to be just right. Not enough and the project is doomed to collapse from unexpected events. However, plans which are too detailed will quickly become out-of-date and will be ignored.

### Time and money
Planning when to do work, and estimating the cost of the resources required for it, must be considered together, except in emergencies.

*Emergencies and urgency*

A project is urgent if the value of completing it faster than normal is greater than the extra cost involved. The designation 'emergency' should be limited to work where the cost is no restraint on using any resources available to work as fast as is physically possible, e.g. in rescue operations to save life. A real emergency is rare.

*A committed project team*

Dispersed project teams tend to correlate with failure; concentrated teams tend to correlate with success. The committed project team should be located where the main risks have to be managed. The separation of people causes a misunderstanding of objectives, communication errors and poor use of expertise and ideas. All people contributing to a project, whether part-time or full-time, should feel that they are committed to a team.

The team should be assembled in time to assess and plan their work and their system of communications. Consultants, suppliers, contractors and others who are to provide goods and services should likewise be appointed in time to mobilize their resources, train and brief staff, and assess and plan their work.

*Representation in decisions*

Success requires the downstream parties to be involved in deciding how to achieve the objectives of a project, and sometimes in setting the objectives themselves. Human systems do not work well if the people who make the initial decisions do not involve those who will be affected later.

*Communications*

The nature of the work on a project changes month by month, and so do the communications needed. The volume and importance of communication can be huge.

Many projects have failed because the communications were poorly organized. A system of communication needs to be planned and monitored, otherwise information comes too late, or goes to the wrong place for decisions to be made. The information then becomes a mere record, and is of little value. These records are then used to allocate blame for problems, rather than to stimulate decisions which will control the problems. The results of informal communications also need to be known and checked, as bad news often travels inaccurately.

*The promoter and the leader*

Every project, large or small, needs a real promoter, a project champion who is committed to its success. Power over the resources needed to deliver a project must be given to one person who is expected to use it to avoid, as well as to manage, problems. In the rest of this book, this person is called 'the project manager'. The project manager may have other jobs, depending upon the size and remoteness of the project. In turn, every sub-project should have its leader with power over the resources it needs.

### Delegation of authority

Inadequate delegation of authority has caused the failure of many projects, particularly where decisions have been restrained by the requirement for approval by people remote from a problem. This has delayed actions and so caused crises, extra costs, and loss of respect and confidence in management.

Many projects have failed because the authority for parts of a large project was delegated to people who did not have the ability and experience to make the necessary decisions. Good delegation requires prior checking that the recipients of delegated authority are equipped to make the decisions delegated to them, and subsequent monitoring of how effectively they are making those decisions. This does not mean making their decisions for them.

### Changes to responsibilities, project scope and plans

Some crises resulting in quick changes to plans are unavoidable during many projects. Drive in solving problems is then very valuable, but failure to think through the decisions made can cause greater problems and loss of confidence in project leaders.

### Control

If the plan for a project is good, the circumstances it assumes materialise and the plan is well communicated, few control decisions and actions are required. Much more is needed if circumstances do change, or if people do not know the plan, or do not understand and accept it.

Control is no substitute for planning. It can waste potentially productive time in reporting and explaining events too late to influence them.

### Reasons for decisions

In project management, every decision leads to the next one and depends upon the one before. The reasons for decisions have to be understood above, below, before and after in order to guide the subsequent decisions. Without this, divergence from the objectives is almost inevitable, and failure is its other name.

Failure to give reasons for decisions and to check that they are accurately understood by their recipients can cause divergent and inconsistent actions. Skill and patience in communication are particularly needed in the rapidly changing relationships typical of the final stages of large projects.

### Using past experience

Success is more likely if technical and project experience from previous projects is drawn upon deliberately and from wherever it is available. Perhaps the frequent failure to do this is another consequence of projects appearing to be unique. It is often easier to say 'this one is different' than to take the trouble to draw experience from previous ones. All projects have similarities and differences. The ability to transfer experience forward by making the appropriate comparisons is one of the hallmarks of a mature applied science.

### Contract strategy

Contract terms should be designed to motivate all parties to try to achieve the objectives of the project and to provide a basis for project management. Contract responsibilities and communications must be clear, and not antagonistic. The terms of the contracts should allocate the risks appropriately between customers, suppliers, contractors and sub-contractors.

### Adapting to external changes

Market conditions, customer's wishes and other circumstances change, and technical problems appear as a project proceeds. Project managers have to be adaptable to these changes, yet able to foresee those which are avoidable and act appropriately.

### Induction, team building and counselling

Success in projects requires people to be brought into a team effectively and rapidly using a deliberate process of induction. Success requires team work to be developed and sustained professionally. It requires people to counsel each other across levels of the organisation, to review performance, to improve, to move sideways when circumstances require, and to respond to difficulties.

### Training

Project management demands intelligence, judgement, energy and persistence. Training cannot create these qualities or substitute for them, but it can greatly help people to learn from their own and other people's experience. After completion, a large project may require retraining of the general management so that they understand, and obtain the full benefit of, its effect on corporate operations.

### Towards perfect projects

The chapters in this book describe the techniques and systems which can be used to apply the lessons of experience. All of them should be considered, but some will be chosen as priorities depending upon the situation and its problems.

All improvements cost effort and money. Cost is often given as a reason not to make a change. In such a situation, the organization should also estimate the cost of not removing a problem.

## References

Wearne, S.H. (1995) *Engineering Project Management*, 1st edn, (Ed. N.J. Smith) Blackwell Science, Oxford.

## Further reading

APM (1998) *Body of Knowledge*, 4th edn, Association of Project Managers, Coventry.

BS 6079 (2000) *Guide to Project Management*, British Standards Institute, London.

Morris, P.W.G. (1997) *The Management of Projects*, 2nd edn, Thomas Telford, London.

O'Connell, F. (1996) *How to Run Successful Projects*, 2nd edn, Prentice Hall, London.

PMI (2000) *A Guide to the Project Management Body of Knowledge*, Project Management Institute, Upper Darby, PA.

Smith, N.J. (1998) *Managing Risk in Construction Projects*, Blackwell Science, London.

Turner, R. (2000) *The Handbook of Project-Based Management*, 3rd edn, Gower, London.

UNIDO (1994) *Manual for the Preparation of Industrial Feasibility Studies*, revised edn, United Nations Industrial Development Organization, New York.

Wearne, S.H., Thompson, P.A. and Barnes, N.M.L. *et al.* (1989) *Control of Engineering Projects*, 2nd edn, Thomas Telford, London.

# Chapter 2
# Value Management

This chapter includes the basic terminology and procedures associated with value management. It then considers the role of value management within the context of a project.

## 2.1 Introduction

Over the past decade, there has been a trend towards applying value techniques at ever earlier stages in a project's life-cycle. The term 'value management' (VM) has become a blanket term that covers all value techniques whether they entail value planning (VP), value engineering (VE) or value analysis (VA).

VM is used by electronics, general engineering, aerospace, automotive and construction industries, and increasingly by service industries. VM techniques have also been successfully applied on all types of construction from buildings to offshore oil and gas platforms, and for all types of clients from private industry to governmental organisations.

## 2.2 Definitions

There are no universally accepted definitions, and a number of different definitions have arisen to describe the same approach or stage of application.

According to the Institution of Civil Engineers (1996):

Value management addresses the value process during the concept, definition, implementation and operation phases of a project. It encompasses a set of systematic and logical procedures and techniques to enhance project value through the life of the facility.

*Value management* (VM) is the title given to the full range of value techniques available, which include those listed below.

*Value planning* (VP) is the title given to value techniques applied during the concept or 'planning' phases of a project. VP is used during the development of the brief to ensure that value is planned into the whole project from its inception. This is achieved by addressing and ranking stakeholders' requirements in order of importance.

*Value engineering* (VE) is the title given to value techniques applied during the design or 'engineering' phases of a project. VE investigates, analyses, compares and selects amongst the various options available those that will meet the value requirements of the stakeholders.

*Value analysis* (VA), or *value reviewing* (VR), is the title given to value techniques applied retrospectively to completed projects to 'analyse' or to audit the project's performance, and to compare a completed, or nearly completed, design or project against predetermined expectations. VA studies are those conducted during the post-construction period and may be part of a post-occupancy evaluation exercise. In addition, the term VA may be applied to the analysis of non-construction-related procedures and processes, such as studies of organisational structure or procurement procedures.

The typical terms for VM studies at different stages of a project, including the various studies which are undertaken, are illustrated in Figure 2.1. VP and VE are applied mainly in the concept and definition phases, and generally end when the design is complete and the construction has started. However, VE can also be applied very effectively during construction to address any problems or opportunities which may arise. The latter typically derive from feedback from the site relating to specific conditions, performance and methods.

*Maximum value* is obtained from a required level of quality as the least cost, the highest level of quality for a given cost, or an optimum compromise between the two. Therefore, VM is the management of a process to obtain maximum value on a scale determined by the client. VM is about enhancing value, and not only about cutting cost (which is often a by-product). The philosophy of VM centres on the identification of the requirements.

Value management involves functional analysis, life-cycle costing, operating in multi-disciplinary groups using the job plan and creativity techniques, and establishing comparative costs in relation to function.

*Functional analysis* is a technique designed to help the appraisal of value by a careful analysis of function (i.e. the fundamental reason why the project element or component exists or is being designed). It explores

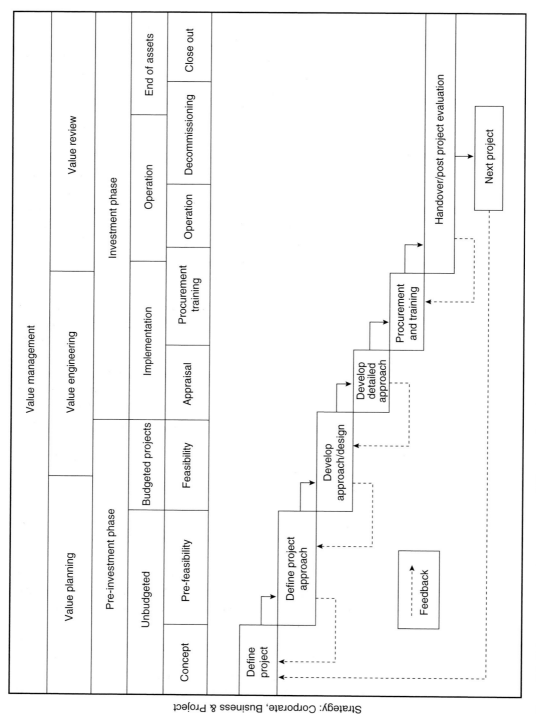

**Figure 2.1** Value management structures at different project phases.

function by asking the initial question 'what does it do?', and then examining how these functions are achieved.

The process is designed to identify alternative, more valuable and/or cost-effective ways to achieve the key functional requirements. Functional analysis is more applicable to the detailed design of specific components (or elements) of a project. Because the function cannot be defined so clearly in the project identification stage, functional analysis is less applicable to this stage.

A job plan is a logical and sequential approach to problem solving which involves the identification and appraisal of a range of options, broken down into their constituent steps and used as the basis of the value management approach. The requirement is defined, different options for resolving the requirement are identified, these options are evaluated and the option(s) with the greatest potential is selected (HM Treasury, Central Unit on Procurement, 1996).

The seven key steps of a job plan are given below.

(1)  Orientation: the identification of what has to be achieved and the key project requirements, priorities and desirable characteristics

(2)  Information: the gathering of relevant data about needs, wants, values, costs, risks, time scale and other project constraints.

(3)  Speculation: the generation of alternative options for the achievement of client needs within stated requirements.

(4)  Evaluation: the evaluation of the alternative options identified in the speculation stage.

(5)  Development: the development of the most promising options and their more detailed appraisal.

(6)  Recommendation for action.

(7)  Implementation and feedback: an examination of how the recommendations were implemented in order to provide lessons for future projects.

As speculation is crucial in the job plan, the quality of ideas generated determine the worth of the approach. Workshops are held where people work together to identify alternative options using idea generating techniques such as brainstorming.

*Life-cycle costing* is an essential part of value management. It is a structured approach used to address all elements of the cost of ownership based on the anticipated life-span of a project.

In the case of a building, some broad categories which can be used for life-cycle costing are given below.

❏ Investment costs: site costs, design fees (architect, quantity surveyor, engineer), legal fees, building costs, tax allowances (capital equipment allowances, capital gains, corporation tax) and development grants.

❏ Energy costs: heating, lighting, air conditioning, lifts, etc.

❏ Non-energy operation and maintenance costs: these include letting fees, maintenance (cleaning and servicing), repair (unplanned replacement of components), caretaker, security and doormen, and insurance rates.

❏ Replacement of components (i.e. planned replacement of components).

❏ Residual or terminal credits. It is necessary to separate the value of the building from the value of the land when determining these credits in the context of a building. Generally, buildings depreciate until they become economically or structurally redundant, whereas land appreciates in value.

A *value management plan* is drawn up which should be flexible, regularly reviewed and updated as the project progresses. This plan should establish:

❏ a series of meetings and interviews;
❏ a series of reviews;
❏ who should attend;
❏ the purpose and timing of the reviews.

## 2.3 Why and when to apply VM

Projects suffer from poor definition because inadequate time and thought are given to them at the earliest stages. This results in cost and time overruns, claims, long-term user dissatisfaction or excessive operating costs. In addition, any project should be initiated only after a careful analysis of need. Therefore, one of the major tasks of stakeholders is to identify at the earliest possible stage the need for, and scope of, any project.

Another basic reason to use VM is because there are almost always elements involved in a project which contribute to poor value. Some of these are:

❏ inadequate time available
❏ habitual thinking/tradition

- conservatism and inertia
- attitudes and influences of stakeholders
- lack of or poor communications
- lack of coordination between the designer and the operator
- lack of a relationship between design and construction methods
- outdated standards or specifications
- absence of state-of-the-art technology
- honest false beliefs/honest misconceptions
- prejudicial thinking
- lack of the necessary experts
- lack of ideas
- unnecessarily restrictive design criteria
- restricted design fee
- temporary decisions that become permanent
- scope of changes for missing items
- lack of essential information

VM is primarily about enhancing value, and not just about cutting cost (although this is often a by-product). VM aims to maximise project value within time and cost constraints without detriment to function, performance, reliability and quality. However, it should be recognised that improving project value sometimes requires extra initial expenditure. The key differences between VM and cost-reduction are that the former is positive, is focused on value rather than cost, is seeking to achieve an optimal balance between time, cost and quality, is structured, auditable and accountable, and is multi-disciplinary. It seeks to maximise the creative potential of all departmental and project participants working together.

When properly organised and executed, VM will help stakeholders to achieve value for money (the desired balance between cost and function which delivers the optimum solution for the stakeholders) for their projects by ensuring that:

- the need for a project is always verified and supported by data;
- project objectives are openly discussed and clearly identified;
- key decisions are rational, explicit and accountable;
- the design evolves within an agreed framework of project objectives;
- alternative options are always considered;
- outline design proposals are carefully evaluated and selected on the basis of defined performance criteria.

Indeed, VM depends fundamentally on whether or not stakeholders can agree on the project objectives from the start.

The key to VM is to involve all the appropriate stakeholders in structured team thinking so that the needs of the four main parties to a project can be accommodated where possible. (HM Treasury, 1996)

The single most critical element in a VM programme is top-level support. (Norton and McElligott 1995)

VM can also provide other important benefits:

❏   improved communication and teamworking;
❏   a shared understanding among key participants;
❏   better quality project definition;
❏   increased innovation;
❏   the elimination of unnecessary cost.

It is apparent that all stakeholders should participate in the process. The only differences are connected with the level and stage of the involvement, and with the party responsible for holding the procedures.

It is paramount that all stakeholders (investors, end-users and others with a real interest in the project outcome, such as the project team, owner, constructors, designers, specialist suppliers) must be involved in the process, especially during the VP and VE stages. In addition, while on larger or more complex projects an independent/external value manager is needed, as well as an external team with the relevant design and technical expertise, on smaller ones VM might be undertaken by the project sponsor's professional adviser, project manager or construction manager. In some cases an external professional must undertake this role. However, when establishing a structure for dealing with value for money on projects, there may be a need for expert assistance, particularly at the 'review' stages.

## 2.4  How to apply VM

There is no single correct approach to VM, but although projects vary, there are a number of stages which are common to all of them. Some of these stages may overlap depending on the type of project. The project sponsor should ensure that a value management plan is drawn up and incorporated into an early draft of the project execution plan (PEP). This plan should establish:

❏   a series of meetings and interviews;
❏   a series of reviews;

❏  who should attend;
❏  the purpose and timing of the reviews.

It should not be a rigid schedule but a flexible plan, regularly reviewed and updated as the project progresses. It is essential that the project sponsor and the value manager prepare for reviews by deciding on the objectives and outputs required, the key participants, and what will be required of them at different stages.

The precise format and timing of reviews will vary according to the particular circumstances and timetable. Too many and the process may be disrupted and delayed, especially at the feasibility stage. Too few and opportunities for improving definition and the effectiveness of the proposals may be lost. Each of these reviews also provide an opportunity to undertake concurrent risk assessments. To exploit the benefits of VM while avoiding unnecessary disruption, there are at least seven obvious 'opportunity points' for reviews which arise on the majority of projects. These are:

❏  during the concept stage, to help identify the need for a project, its key objectives and its constraints;
❏  during the pre-feasibility stage, to evaluate the broad project approach/outline design;
❏  during the scheme design (feasibility stage), to evaluate developing design proposals;
❏  during the detailed design (appraisal stage), to review and evaluate key design decisions;
❏  during construction/implementation, to reduce costs or improve buildability or functionality;
❏  during commissioning/operation, to remedy possible malfunctions or deficiencies;
❏  during decommissioning/end of assets, to learn lessons for future projects.

## 2.5  Reviews

All reviews should be structured to follow the job plan. Issues of buildability, safety, operation and maintenance should be considered during all VM reviews and evaluation options.

The first review should cover the items listed below.

(1)  List all objectives identified by stakeholders.
(2)  Establish a hierarchy of objectives by ranking them in order of

priority. It is important to stress that the aim is to produce a priority listing, not simply to drop lesser priorities. Reducing the list runs the risk of having to reintroduce priorities at a later stage, with all the associated detrimental impacts on cost, time and quality. VM aims to eradicate the need for late changes, it should not encourage them.

(3)   Identify broad approaches to achieving objectives by brainstorming.
(4)   Appraise the feasibility of options (reject/abandon, delay/postpone).
(5)   Identify potentially valuable options.
(6)   Consider, and preferably recommend, the most promising option for further development.

The first review should show the following results.

(1)   Confirmation that the project is needed.
(2)   A description of the project, i.e. what has to be done to satisfy the objectives and priorities.
(3)   A statement of the primary objective.
(4)   A 'hierarchy' of project priorities.
(5)   A favoured option(s) for further development.
(6)   A decision to proceed.
(7)   A decision to reject/abandon or postpone/delay, if necessary.

This balanced statement of needs, objectives and priorities, agreed by all stakeholders, helps the project sponsor to produce the project brief.
   The second review should cover the items listed below.

(1)   Review the validity of the hierarchy of objectives with stakeholders and agree modifications.
(2)   Evaluate the feasibility of the options identified.
(3)   Examine the most promising option to see if it can be improved further.
(4)   Develop an agreed recommendation about the most valuable option which can form the basis of an agreed project brief.
(5)   Produce a programme for developing the project.

The second review should show the following results.

(1)   A clear statement of the processes to be provided and/or accommodated.

(2)   A preferred outline design proposal.
(3)   The basis of a case for the continuation of design development.

During the third phase/feasibility stage, some 10–30% of the design work will be completed. That is why VE techniques are also part of this stage which should cover the following points.

(1)   Review the project requirements and the hierarchy of objectives agreed at the last review.
(2)   Check that the key design decisions taken since the last review remain relevant to the hierarchy of objectives and priorities.
(3)   Review key decisions against the project brief by brainstorming to identify ways of improving the design proposals outlined to date and to identify options.
(4)   Evaluate options in order to identify the most valuable one.
(5)   Develop the most valuable option to enhance value, focusing on and resolving any perceived problems.
(6)   Agree on a statement of the option to be taken forward, and agree on a plan for the continued development of the design.

The third review should show the following results.

(1)   A thorough evaluation of the sketch design.
(2)   Clear recommendations for the finalisation of the sketch design.
(3)   The basis of a submission for final approval to implement, abandon or postpone the project.

VE is aimed at finding the engineering, architectural and technical solutions to help translate the design scheme selected by VP into a detailed design which provides the best value, by analysing, evaluating and recommending the proposals of the constructor and addressing problems that may emerge during construction.

By reviewing design proposals in this way, the value team will seek answers to the following questions to help determine value and function.

❏   What is it (i.e. the purpose of the project or element)?
❏   What does it do?
❏   What does it cost?
❏   How valuable is it?
❏   What else could do the job?
❏   What will that cost?

The fourth review should show the following results.

(1)   Promotion of a continuous VM approach throughout the design process.
(2)   Finalisation of the original and proposed designs and the basis for the changes, according to the findings of the previous review.
(3)   A description of the value proposals, and an explanation of the advantages and disadvantages of each in terms of estimated savings, capital, operating and life-cycle costs, and improvements in reliability, maintenance or operation.
(4)   A prediction of the potential costs and savings, the redesign fee and the time associated with the recommended changes.
(5)   A timetable for decisions of the owner, implementation costs, procedures and any problems (such as delays) which may reduce benefits.

The fifth review should show the following results.

(1)   Promotion of a continuous VM approach throughout the construction process.
(2)   An assessment and evaluation of the contractors' proposed changes.
(3)   An investigation and verification of the feasibility of significant changes and the cost savings claimed, as well as the implications of including them in the programme.
(4)   Some forward-looking, practical recommendations for improvements which can be implemented immediately.
(5)   Check that any risks to the project are being managed.

The sixth review should show the following results.

(1)   A measure of the success of the project in achieving its planned objectives.
(2)   Identification of the reasons for any problems that have arisen.
(3)   A decision on what remedial actions should be taken.
(4)   An assessment of whether the objectives of the users/customers have been met. If those objectives changed, or were expected to change during the course of the project, to assess whether those changes were accommodated.
(5)   A check that any outstanding work, including defects, have been remedied.
(6)   A record of the lessons which have been learnt which could improve performance on subsequent or continuing projects.

Practitioners and users of VM must obtain feedback on its success, since feedback influences the results by raising the following questions.

❑ Have good ideas emerged?
❑ Were any adopted?
❑ Were they implemented?
❑ Did the expected value improvement result?
❑ If not, why not?

Factors which will be assessed include the judgement, involvement, support, application, dedication, foot-dragging, approval process, systems appropriateness, use, effectiveness and management of the change process of the stakeholders.

Actions that can emerge from the feedback include:

❑ a change of personnel;
❑ a change of approach;
❑ a change of system;
❑ a rerun of the exercise.

## 2.6  Procedures and techniques

There are many procedures and techniques available within VM for the value team to use as they see fit, whether applied former or intuitively. Typical techniques and procedures include:

❑ information gathering
❑ cost analysis
❑ life-cycle costing
❑ Pareto's law
❑ basic and secondary functions
❑ cost and worth
❑ FAST diagramming (function analysis)
❑ creative thinking and brainstorming
❑ criteria weighting
❑ value tree
❑ checklists/attribute listing
❑ analysis and ranking of alternatives

## 2.7  Benefits of value management

Properly organised and executed VM provides a structured basis for both the appraisal and development of a project, and results in many benefits to that project. There follows a very brief list of such benefits.

- Provides a forum for all concerned parties in a project development.
- Develops a shared understanding among key participants.
- Provides an authoritative review of the entire project, not just a few elements.
- Identifies project constraints, issues and problems which might not otherwise have been identified.
- Identifies and prioritises the key objectives of a project.
- Improves the quality of definition.
- Identifies and evaluates the means of meeting needs and objectives.
- Deals with the life cycle, not just initial costs.
- Usually results in remedying project deficiencies and omissions, and superfluous items.
- Ensures all aspects of the design are the most effective for their purpose.
- Identifies and eliminates unnecessary costs.
- Provides a means to identify and incorporate project enhancements.
- Provides a priority framework against which future potential changes can be judged.
- Crystallises an organisation's brief priorities.
- Maintains a strategic focus on the organisation's needs during the development and implementation of a project.
- Provides management with the information it needs to make informed decisions.
- Permits a large return on a minimal investment.
- Promotes innovation.

## 2.8  Summary

The key features of the value process and the application of VM to it have been described, and the importance of value planning, teamwork and perseverance emphasised. The incentives and benefits to all stakeholders have been identified and discussed. These are underpinned by three particular aspects: the independence of the value manager clearly to establish the stakeholders' value criteria; the planned application of team brainstorming; the inclusion of appropriate enabling clauses in contracts and agreements.

The factors needed to ensure success of VM include:

❏  a systematic approach;
❏  an integrated team environment;
❏  the establishment of value criteria;
❏  focusing on the function;
❏  the facilitation of creativity as a separate stage;
❏  consideration of a project on a life-cycle cost basis;
❏  a collaborative and non-confrontational working environment;
❏  the generation of records and an audit trail.

VM must have comprehensive top management understanding and support, and an enthusiastic, sustained and innovative approach.

## References

HM Treasury (1996) Central Unit on Procurement Guidance Note 54, *Value Management*, HMSO, London.
Institution of Civil Engineers (1996) *Creating Value in Engineering Projects*, Thomas Telford, London.
Norton, B.R. and McElligott, W.C. (1995) *Value Management in Construction*, Macmillan, London.

## Further reading

Connaughton, J.N. and Green, S.D. (1996) *Value Management in Construction: A Client's Guide*, CIRIA, London.
Webb, A. (1994) *Managing Innovative Projects*, Chapman & Hall, London.

# Chapter 3

# Project Appraisal and Risk Management

Project appraisal, sometimes referred to as feasibility study, is an important stage in the evolution of a project. It is important to consider alternatives, and identify and assess risks at a time when data is uncertain or unavailable. This chapter outlines the stages of a project, and describes risk management techniques in detail.

## 3.1 Initiation

An individual project, however significant and potentially beneficial to the promoting organisation, will only constitute part of a corporate business. It is also likely that, in the early stages of the project cycle, several alternative projects will be competing for the available resources, particularly finance. The progress of any project will therefore be subject to investment decisions by the parent organisation before the project is allowed to proceed.

In most engineering projects, the rate of expenditure changes dramatically as the project moves from the early stages of studies and evaluations, which consume mainly human expertise and analytical skills, to the design, manufacture and construction of a physical facility. A typical investment curve is shown in Figure 3.1, and indicates that considerable cost will be incurred before any benefit accrues to the promoter from use of the completed project (or asset).

When considering the investment curve, the life-cycle of the project can be divided into three major phases: the appraisal and implementation of the project, followed by the operation of the completed facility. The precise shape of the curve will be influenced by the nature of the project, by external factors such as statutory approvals, and also by the project's objectives. In the public sector, appraisal may extend over many years and be subjected to several intermediate decisions to proceed. In a commercial situation, the need for early entry into a competitive market may outweigh all other considerations.

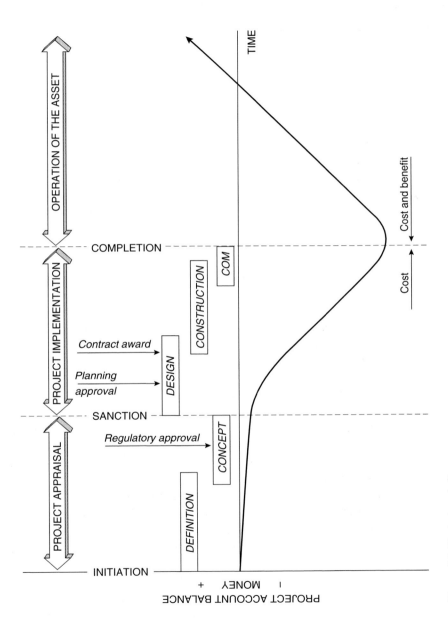

**Figure 3.1** Project cash flow.

Two important factors emerge when studying the investment curve:

❏ That interest payments compounded over the entire period when the project account is in deficit (below the axis in Figure 3.1) will form a significant element of project costs.
❏ That the investor will not derive any benefit until the project is completed and in use.

It is likely that at least two independent and formal investment decisions will be necessary; they are labelled 'initiation' or 'viability', and 'sanction' in Figure 3.1. The first signifies that the ill-defined idea evolved from research and studies of demand is perceived to offer sufficient potential benefit to warrant the allocation of a specific project budget for further studies, and the development of the project concept. The subsequent sanction decision signifies acceptance or rejection of these detailed proposals. If the decision is positive, the organisation will then proceed with the major part of the investment in the expectation of deriving some predicted benefit when the project is completed.

All these estimates and predications, which frequently extend over a period of many years, will generate different degrees of uncertainty. The sanction decision therefore implies that the investor is prepared to take the risk.

## 3.2 Sanction

When a project is sanctioned, the investing organisation is committing itself to major expenditure and is assuming the associated risks. This is the key decision in the life-cycle of the project. In order to make a well-researched decision, the promoter will require information under the headings listed below.

(1) *Clear objectives.* The promoter's objectives in pursuing this investment must be clearly stated and agreed by senior management early in the appraisal phase, for all that follows is directed at achieving these objectives in the most effective manner. The primary objectives of quality, time and cost may well conflict, and it is particularly important that the project team know whether a minimum time for completion or minimum cost is the priority. These are rarely compatible, and this requirement will greatly influence both the appraisal and the implementation of the project.

(2) *Market intelligence.* This relates to the commercial environment in

which the project will be developed and later operated. It is necessary to study and predict trends in the market and the economy, and anticipate technological developments and the actions of competitors.

(3) *Realistic estimates/predictions*. It is easy to be over optimistic when promoting a new project. Estimates and predictions made during appraisal will extend over the whole life-cycle of the implementation and operation of the project. Consequently, single-figure estimates are likely to be misleading, and due allowance for uncertainty and exclusions should be included.

(4) *Assessment of risk*. A thorough study of the uncertainties associated with the investment will help to establish confidence in the estimate, and to allocate appropriate contingency funds. More importantly at this early stage of project development, it will highlight areas where more information is needed, and will frequently generate imaginative responses to potential problems, thereby reducing risk.

(5) *Project execution plan*. This should give guidance on the most effective way to implement the project and to achieve its objectives, taking account of all constraints and risks. Ideally, this plan will define the likely contract strategy, and include a programme showing the timing of key decisions and the award of contracts.

It is widely held that the success of a venture is largely dependent on the effort expended during the appraisal preceding sanction. However, there is conflict between the desire to gain more information and thereby reduce uncertainty, the need to minimise the period of investment, and the knowledge that expenditure on appraisal will have to be written off if the project is not sanctioned.

Expenditure on the appraisal of major engineering projects rarely exceeds 10% of the capital cost of the project. However, the outcome of the appraisal, as defined in the concept and brief accepted at sanction, will freeze 80% of the cost. The opportunity to reduce the cost during the subsequent implementation phase is relatively small, as shown in Figure 3.2.

## 3.3 Project appraisal and selection

Project appraisal is a process of investigation, review and evaluation undertaken as the project, or alternative concepts of the project, are defined. This study is designed to assist the promoter to make informed and rational choices concerning the nature and scale of investment in the

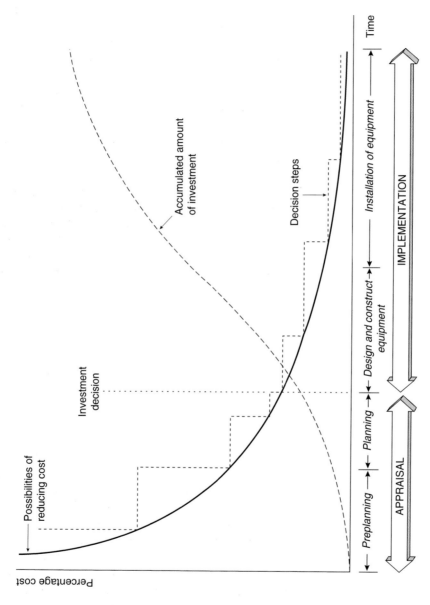

**Figure 3.2** Graph of the percentage cost against time taken, showing how the important decisions for any project are made at the start of that project.

project, and to provide the brief for subsequent implementation. The core of the process is an economic evaluation, based on a cash flow analysis of all costs and benefits which can be valued in money terms, which contributes to a broader assessment called cost–benefit analysis. A feasibility study may form part of the appraisal.

The appraisal is likely to be a cyclic process repeated as new ideas are developed, additional information received and uncertainty reduced, until the promoter is able to make the critical decision to sanction implementation of the project and commit to the investment in anticipation of the predicted return. It is important to realise that if the results of the appraisal are unfavourable, this is the time to defer further work or abandon the project. The consequences of an inadequate or unrealistic appraisal can be expensive.

Ideally, all alternative concepts and ways of achieving the project objectives should be considered. The resulting proposal prepared for sanction must define the major parameters of the project: the location, the technology to be used, the size of the facility, and the sources of finance and raw materials, together with forecasts of the market and predictions of the cost–benefit of the investment. There is usually an alternative way to utilise resources, especially money, and this is capable of being quantified, however roughly.

Investment decisions may be constrained by non-monetary factors such as:

❑ organisational policy, strategy and objectives;
❑ the availability of resources such as manpower, management or technology.

*Programme*

It will be necessary to decide when is the best time to start the project based on the previous considerations. Normally, this means as soon as possible, because no profit can be made until the project is completed. Indeed, it may be that market conditions or other commitments impose a programme deadline; for example, a customer will not buy your product unless it is available by mid-2002, when the customer's processing factory will be ready. In inflationary times, it is doubly important to complete a project as soon as possible because of the adverse relationship between time and money. The cost of a project will double in 7.25 years at a rate of inflation of 10%.

It will therefore be necessary to determine the duration of the appraisal, design and construction phases

❏   so that the operation date can be determined,
❏   so that project costs can be determined,
❏   and so that the promoter's liabilities can be assessed and checked for viability. It may well be that the promoter's cash availability defines the speed at which the project can proceed.

The importance of time should be recognised throughout the appraisal. Many costs are time-related, and would be extended by any delay. The programme must therefore be realistic, and its significance taken fully into account, when determining the objectives of the project.

### Risk and uncertainty

The greatest degree of uncertainty about the future is encountered early in the life of a new project. Decisions taken during the appraisal stage have a very large impact on final cost, duration and benefits. The extent and effects of change are frequently underestimated during this phase, although these are often considerable, particularly in developing countries and remote locations. The overriding conclusion drawn from recent research is that all parties involved in construction projects would benefit greatly from reductions in uncertainty prior to financial commitment.

At the appraisal stage, the engineering and project management input will normally concentrate on providing:

❏   a realistic estimate of capital and running costs;
❏   a realistic time-scale and programmes for project implementation;
❏   appropriate specifications for performance standards.

At appraisal, the level of project definition is likely to be low, and therefore the risk response should be characterised by a broad-brush approach. It is recommended that efforts should concentrate on:

❏   seeking solutions which avoid/reduce risk;
❏   considering whether the extent or nature of the major risks are such that the normal transfer routes may be unavailable or particularly expensive;
❏   outlining any special treatment which may need to be considered for risk transfer, for example for insurance or unconventional contractual arrangements;

❏ setting realistic contingencies, and estimating tolerances consistent with the objective of preparing the best estimate of anticipated total project cost;

❏ identifying comparative differences in the riskiness of alternative project schemes.

Engineering/project managers will usually have less responsibility for identifying the revenues and benefits from the project; this is usually the function of marketing or development planning departments. The involvement of project managers in the planning team is recommended, as the appraisal is essentially a multi-disciplinary brainstorming exercise through which the promoter seeks to evaluate all alternative ways of achieving these objectives.

For many projects this assessment is complex, as not all the benefits/disbenefits will be quantifiable in monetary terms. For others, it may be necessary to consider the development in the context of several different scenarios (or views of the future). In all cases, the predictions are concerned with the future needs of the customer or community. They must span the overall period of development and operation of the project, which is likely to range from a minimum of 8 or 10 years for a plant manufacturing consumer products, to 30 years for a power station and much longer for public works projects. Phasing of the development should always be considered.

Even at this early stage of project definition, maintenance policies and requirements should be stated as these will affect both design and cost. Special emphasis should be given to future maintenance during the appraisal of projects in developing areas. The cost of dismantling or decommissioning may also be significant, but is frequently conveniently ignored.

## 3.4 Project evaluation

The process of economic evaluation and the extent of uncertainty associated with project development require the use of a range of financial criteria for the quantification and ranking of the alternatives. These will normally include discounting techniques, but care must be taken when interpreting the results for projects of long duration.

### Cost–benefit analysis

In most engineering projects, factors other than money must be taken into account. If a dam is built, we might drown a historic monument, reduce the likelihood of loss of life due to flooding, increase the growth of new industry because of the reduced risk, and so on. Cost–benefit analysis provides a logical framework for evaluating alternative courses of action when a number of factors are highly conjectural in nature. If the evaluation is confined to purely financial considerations, it fails to recognise the overall social objective, which is to produce the greatest possible benefit for a given cost.

At its heart lies the recognition that no factor should be ignored because it is difficult, or even impossible, to quantify in monetary terms. Methods are available to express, for instance, the value of recreational facilities, and although it may not be possible to put a figure on the value of human life, it is surely not something we can afford to ignore.

Essentially, cost–benefit analysis must take into account all the factors which influence either the benefits or the cost of a project. Imagination must be used to assign monetary values to what at first sight might appear to be intangibles. Even factors to which no monetary value can be assigned must be taken into consideration. The analysis should be applied to projects of roughly similar size and patterns of cash flow. Those with a higher cost–benefit ratio will be preferred. The maximum net benefit ratio will be marginally greater than the next most favoured project. The scope of the secondary benefits to be taken into account frequently depend on the viewpoint of the analyst.

In comparing alternatives, it is obvious that each project should be designed at the minimum cost which will allow the fulfilment of the objectives, including the appropriate quality, level of performance and provisions for safety. Perhaps more importantly, the viewpoint from which each project is assessed plays a critical part in properly assessing both the benefits and cost which should be attributed to a project. For instance, if a private electricity board wishes to develop a hydroelectric power station, it may derive no benefit from the coincidental provision of additional public recreational facilities, which therefore cannot enter into its cost–benefit analysis. However, a public-sector owner could quite properly include the recreational benefits in its cost–benefit analysis. Again, as far as the private developer is concerned, the cost of labour is equal to the market rate of remuneration, no matter what the unemployment level. For the public developer, however, in times of high unemployment, the economic cost of labour may be nil, since the use of labour in this project does not preclude the use of other labour for other purposes.

## 3.5  Engineering risk

An essential aspect of project appraisal is the reduction of risk to a level which is acceptable to the investor. This process starts with a realistic assessment of all uncertainties associated with the data and predictions generated during appraisal. Many of the uncertainties will involve a possible range of outcomes, i.e. it could be better or worse than predicted. Risks arise from uncertainty, and are generally interpreted as factors which have an adverse effect on the achievement of the project objectives.

It is helpful to try to categorise the risks associated with projects both as a guide to identification, and to facilitate the selection of the most appropriate risk-management strategy. One method is to separate the more general risks which might influence a project, but which may be outside the control of the parties to the project, from the risks associated with key project elements; these are referred to as global and elemental risks, respectively.

Global risks may be capable of being influenced by governments, and can be sub-divided into four sections: political, legal, commercial and environmental risk. Political risk would include events such as a public inquiry, approvals, regulation of competition and exclusivity, whereas changes in statue law, regulations and directives would all be legal risks. Commercial risks can include the wider aspects of demand and supply, recession and boom, social acceptability and consumer resistance. Environmental risks are easier to identify, but as global risks they are more specifically to do with changes in standards, in external pressure, and in environmental consents.

The elemental risks are those associated with elements of the project, namely implementation risks and operation risks. For some projects there will also be financial risks and revenue risks. These risks are more likely to be controllable or manageable by the parties to the project. Typical examples of these risks are given in Tables 3.1–3.4.

## 3.6  Risk management

The logical process of risk management may be defined as:

❏  identification of risks/uncertainties;
❏  analysis of the implications (individual and collective);
❏  response to minimise risk;
❏  management of residual risk.

**Table 3.1** Implementation risks.

| Risk category | Description |
| --- | --- |
| Physical | Natural: pestilence and disease, ground conditions, adverse weather conditions, physical obstructions |
| Construction | Availability of plant and resources, industrial relations, quality, workmanship, damage, construction period, delay, construction programme, construction techniques, milestones, failure to complete, type of construction contracts, cost of construction, insurances, bonds, access, insolvency |
| Design | Incomplete design, design life, availability of information, meeting specifications and standards, changes in design during construction, design life, competition of design |
| Technology | New technology, provision for change in existing technology, development costs |

**Table 3.2** Operational risks.

| Risk category | Description |
| --- | --- |
| Operation | Operating conditions, raw materials, supply, power, distribution of offtake, plant performance, operating plant, interruption to operation due to damage or neglect, consumables, operating methods, resources to operate new and existing facilities, type of O&M contract, reduced output, guarantees, underestimation of operating costs, licences |
| Maintenance | Availability of spares, resources, sufficient time for major maintenance, compatibility with associated facilities, warranties |
| Training | Cost and levels of training, translation of manuals, calibre and availability of personnel, training of principal personnel after transfer |

**Table 3.3** Financial risks.

| Risk category | Description |
| --- | --- |
| Interest | Type of rate (fixed, floating or capped), changes in interest rate, existing rates |
| Payback | Loan period, fixed payments, cash-flow milestones, discount rates, rate of return, scheduling of payments, financial engineering |
| Loan | Type and source of loan, availability of loan, cost of servicing loan, default by lender, standby loan facility, debt/equity ratio, holding period, existing debt, covenants, financial instruments |
| Equity | Institutional support, take-up of shares, type of equity offered |
| Dividends | Time and amounts of dividend payments |
| Currencies | Currencies of loan, ratio of local/base currencies |

**Table 3.4** Revenue risks.

| Risk category | Description |
| --- | --- |
| Demand | Accuracy of demand and growth data, ability to meet increase in demand, demand over concession period, demand associated with existing facilities |
| Toll | Market-led or contract-led revenue, shadow tolls, toll level, currencies of revenue, tariff variation formula, regulated tolls, take and/or pay payments |
| Developments | Changes in revenue streams from developments during concession period |

Risk management can be considered as an essential part of a continuous and structured project planning cycle. Risk management:

❏ requires that you accept that uncertainty exists;
❏ generates a structured response to risk in terms of alternative plans, solutions and contingencies;
❏ is a thinking process requiring imagination and ingenuity;
❏ generates a realistic (and sometimes different) attitude in project staff by preparing them for risk events rather than letting them be taken by surprise when they arise.

If uncertainty is managed realistically, the process will:

❏ improve project planning by prompting 'what if' questions;
❏ generate imaginative responses;
❏ give greater confidence in estimates;
❏ encourage the provision of appropriate contingencies and consideration of how they should be managed.

Risk management should impose a discipline on those contributing to the project, both internally and on customers and contractors. By predicting the consequences of a delayed decision, failure to meet a deadline, or a changed requirement, appropriate incentives/penalties can be devised. The use of range estimates will generate a flexible plan in which the allocation of resources and the use of contingencies is regulated.

### Risk reduction

Risk reduction includes:

❏ obtaining additional information;
❏ performing additional tests/simulations;
❏ allocating additional resources;
❏ improving communication and managing organisational interfaces;
❏ allocating risk to the party best able to control or manage that risk.

Market risk may frequently be reduced by staging the development of the project. All the actions listed above will incur additional cost in the early stages of project development.

*The role of people*

All the risks described above may be aggravated by the inadequate performance of individuals and organisations contributing to the project.

Control is exercised by and through people. As project managers will need to delegate, they must have confidence in the members of the project or contract team, and ideally should be involved in their selection.

Involve staff in risk management in order to utilise their ideas and to generate motivation and commitment. The roles, constraints and procedures must be clear, concise and understood by everyone with responsibility.

## 3.7  Risk and uncertainty management

Recently, much attention has been given to the prospect of considering both the risks and the opportunities associated with a particular project. This is sometimes known as uncertainty management. The evaluation of both risk and opportunity are affected by uncertainty. However, it is important to note that these are different processes, requiring a different mind-set and different data, and cannot easily be integrated at a detailed level for any project.

In the UK, the Turnbull Report requires all companies to include in their annual statement of accounts a guide showing how risks and opportunities are monitored and managed. This has tended to lead companies to produce simple risk and opportunity matrices, with various levels of impact. This has its own dangers, and should be an addition to detailed risk-management and value-management studies being undertaken as and when required.

No standard approach has been developed for the industrial application of uncertainty management, but this is another example of the continuing evolution of project management theory and practice, as described in Chapter 20.

## Further Reading

Misham, E.J. (1988) *Cost–Benefit Analysis*, Unwin, London.
Perry, J.G. and Thompson, P.A. (1992) *Engineering Construction Risks*, Thomas Telford, London.
Smith, N.J. (1998) *Managing Risk in Construction*, Blackwell Science, Oxford.

# Chapter 4
# Project Management and Quality

Quality is an important concern for all business organisations. Although the management of quality is perceived to be a relatively recent concept, the Institute of Quality Assurance was established in 1919. Most business organisations now require potential partners, suppliers and vendor organisations to operate a quality system. For many years, project managers have used procedures for certain types of projects which can be adopted for the execution and integration of a quality system. This chapter describes how project management can be used effectively to develop, support and administer project-specific quality systems. To ensure safety and to limit the consequences of any damage, total quality management is required.

## 4.1 Definitions

There are many definitions of quality, which are usually strongly influenced by a particular industrial sector. Several definitions are listed below.

❑ Quality is the ability to meet market and customer expectations, needs and requirements.
❑ Quality is the supplying of goods which do not come back, to customers who do.
❑ Quality means in conformance with user requirements.
❑ Quality means fitness for use.

Essentially, quality can be described as needing the six 'C's:

❑ confidence
❑ control
❑ consistency

- cost effectiveness
- commitment
- communication

For the purposes of this chapter, the author suggests that quality topics should be defined as those currently stated by the Association for Project Management, which are given below.

- Quality is ensuring that required standards of performance are attained. Three different stages of quality management are typically encountered.
- Quality assurance (QA) defines the procedures and documentation requirements to establish a predetermined level of performance.
- Quality control (QC) is the process of measuring that a predefined level of performance has been achieved.
- Total quality management (TQM) is a much broader and ambitious process involving:
  - identifying what (standards, performance, requirements) the customer really wants;
  - defining the mission of the organisation;
  - involving all personnel in identifying how these two aims could be better achieved;
  - designing ways in which performance could be approved;
  - measuring, throughout the total production process, how well performance meets the required standards;
  - analysing continually how performance can be further improved.

Currently, most project organisations require potential partner, supplier or vendor organisations to operate a quality system. The signs are that this trend is increasing, and those businesses that do not have a quality system will lose business opportunities. Suppliers, distributors and others providing products and services are increasingly being held responsible for any damage caused by the product to persons or property. This is known as product liability.

In the UK, the quality assurance standard BS5750 was produced in 1979, and the International Standards Organisation produced ISO 9000–9004 in 1987. Table 4.1 shows the quality system elements from the ISO 9000 standards and their requirements. There is now a compatible European equivalent, EN29000. These documents provide guidance on the preparation and implementation of quality systems. Organisations achieving the required standards are certified for given periods of time.

**Table 4.1** Quality assurance, ISO 90001.

Management responsibility
Quality system
Contract review
Design control
Document control
Purchasing
Purchaser-supplied products
Product identification and traceability
Process control
Inspection and testing
Inspecting, measuring and testing equipment
Inspection and test status
Control of non-conforming product
Corrective action
Handling, storage, packaging and delivery
Quality records
Internal quality audits
Training
Servicing
Statistical techniques

## 4.2  Quality systems

A project quality system incorporates all stages of the project from conception to operation; sometimes until decommissioning. By focusing on the early feasibility and design stages, quality and cost requirements may be identified. In order to achieve the desired quality without unnecessary costs, a project manager must pursue an efficient system of coordinating the project activities.

The quality system should ensure that:

❑ the quality products and services always meet the expressed or implied requirements of the customer;
❑ the company management knows that quality is achieved in a systematic way;
❑ the customer feels confident about the quality of goods or services supplied and the method by which they are achieved.

The quality system must be adjusted to suit the on-going operation as well as the final product. It must be designed so that the emphasis is put on preventive actions, at the same time allowing the project manager to correct any mistakes that do occur during the life-cycle of the project.

The quality system should be based on the following activities:

❑ planning
❑ execution
❑ checking
❑ action

All activities and tasks affecting the quality of a project require planning to achieve a satisfactory result. Execution should be based on relevant expertise and resources, and results must be checked. Checking must be followed by action. Defective products or sections of work must be removed. All information gained must be analysed and recorded in order to prevent the same defects from appearing again. Continuous upgrading of the quality system must take place. This interrelationship is shown by the 'Deming Circle' in Figure 4.1.

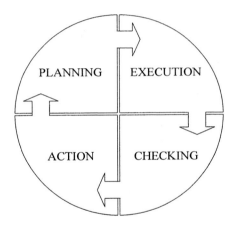

**Figure 4.1** The Deming circle.

Certain common rules are necessary to coordinate and direct a project's operation in order to achieve the quality objectives. These must be made known to all members of the company and the project team. These rules operate in a hierarchy illustrated in Figure 4.2, and each step is described below.

*Quality policy* is the main guide to the company's approach to quality matters. It clarifies the overall principles if the company's attitude towards, and handling of, quality management. The quality policy must not make any statement that will not have the backing and resources necessary for its achievement.

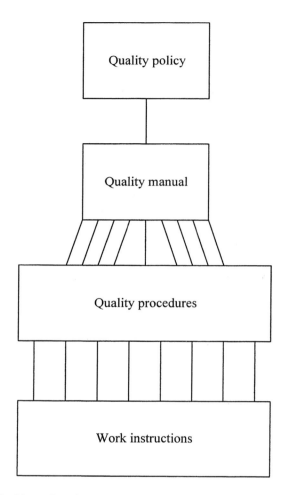

**Figure 4.2** The hierarchy of quality procedure.

*Quality principles* give more specialised guidelines as to the specific project phases. In order to ensure that staff are aware of the methods of working and their responsibilities, it is essential that principles and procedures are documented. This is done in the form of a quality manual.

*Quality procedures* define the patterns and sequences of actions to be taken, and coordinate the construction processes which are important to quality matters. They distribute the responsibility between different functions, i.e. departments within the company and members of the project team.

*Working instructions* give detailed information as to the execution of the various stages of construction related to the project.

The aim of the quality manual is to aid the company and project team in directing their quality management. It is becoming increasingly common for customers to demand proof from suppliers of a documented quality system; the manual will also fulfil this function. It should provide an adequate description of the quality management system, while serving as a permanent reference in the implementation and maintenance of that system. The quality manual is also used as means to educate new employees by identifying the routines and responsibilities of all other employees within the company or specific to a project.

The company management and the project manager should periodically review the quality actions. A tool for this is the quality audit. Quality audits are similar to other in-house audits, but they concentrate on quality aspects. A quality audit is a systematic and independent examination to determine whether quality activities and results comply with planned arrangements, and whether these arrangements are implemented effectively and are suitable to achieve the objectives. It can be directed towards goals (goal audit), systems (system audit) or structure and practicality (operation audit).

## 4.3 Implementation

Implementing a project-specific quality system involves the following actions:

- ❑  determine the size of the project, the project team and its location;
- ❑  identify the status of existing documentation, and determine whether or not this is compatible with the proposed project activities;
- ❑  determine the range of project activities within the specific project;
- ❑  identify the level of activities within the project and the method of work allocation;
- ❑  determine the responsibilities and commitment of the project management team.

The structure of documentation within a quality system should take account of structure in order to avoid overlapping and duplication, design in order to provide a uniform simple format, distribution in order to devise a manageable system that can easily be updated, and the facility for documentation changes.

A documentation system must be able to develop so that it clearly reflects the activities and methods contained in the project. Sections of the system will need to be rapidly updated and reissued to keep abreast of demands, and any changes in techniques and procedures.

All members of the project team should have a job description. The job description should define the following parameters: designation within the project, reporting procedures, key tasks and responsibilities, relationship with other members of the project team, reference to work instructions, specific performance measurement where applicable, and specified minimum training required.

It is recognised that training is one of the cornerstones of a quality system. Certain training is mandatory for employees and, wherever practical, should be carried out prior to any work responsibility being assigned. The following areas must be covered by employees involved in a project: the company introduction package and specific training regarding health and safety at work and the project quality system.

The advantages to the project manager of developing and introducing a quality system are:

❑ that profits would be increased through better management control systems;
❑ quality levels would be uniform throughout the project;
❑ safer deliveries and storage of materials;
❑ costs should be reduced through higher productivity, and fewer defects and changes;
❑ capital utilisation should be improved through faster and safer throughput;
❑ satisfied clients and/or promoters.

Quality costs, but lack of quality costs even more. Today's society depends heavily on reliable products and services. The consequences of failure are costly, both to society as a whole and to the project, incurring additional expense and inconvenience.

## 4.4 Quality-related costs

The project manager is primarily concerned with controlling time and costs, and with the inclusion of a quality system they should be in a position to control quality as well. The need to identify the costs of achieving quality is not new, although the practice is not widespread. In this way, a monetary value can be placed on wasted effort and on the costs of correcting these errors, and hence the work on the project can be redirected to optimise benefit.

Quality costs can be either operating quality costs or external assurance quality costs.

Operating costs are those incurred by a project in order to attain and ensure specified quality levels. These include those listed below.

### Prevention and appraisal project costs (or investments)

- ❑ Prevention costs, i.e. the costs of efforts to prevent failures, include:
  - ■ design reviews,
  - ■ quality and reliability training,
  - ■ vendor quality planning,
  - ■ audits,
  - ■ construction and installation prevention activities,
  - ■ product qualification,
  - ■ quality engineering.
- ❑ Appraisal costs, i.e. the costs of testing, inspection and examination to assess whether a specified quality is being maintained, include:
  - ■ tests and inspections,
  - ■ maintenance and calibration,
  - ■ test equipment depreciation,
  - ■ line-quality engineering,
  - ■ installation testing and commissioning.

### Failure costs (or losses)

- ❑ Internal failure costs resulting from a product or a service failing to meet the performance specification prior to delivery (e.g. product service, warranties and returns, direct costs and allowances, product recall costs, liability costs). These include:
  - ■ design changes,
  - ■ vendor rejects,
  - ■ rework,
  - ■ scrap and material renovation,
  - ■ warranties,
  - ■ commissioning failures,
  - ■ fault-finding in tests.

External assurance quality costs are described below.

- ❑ External assurance quality costs are those relating to the demonstration and proof required by promoters as evidence of work done, including particular and additional quality assurance provisions, procedures, data and documentation.

- External quality assurance costs should be lower than operating costs since most quality costs will have been incurred before the involvement of external quality assurance personnel. Ideally, external assurance personnel should be involved with both the project and the company.

Only by knowing where costs are incurred, and their order of magnitude, can project managers monitor and control them. Quality costs must be collected and recorded separately, otherwise they become absorbed and concealed in numerous other overheads. Regular financial reports are vital if there is to be management visibility into quality-related costs.

There is a balance between each of these three types of quality cost and their relationship with improvements in quality. As would be expected, each of the three types of quality cost reduce as quality improves, but the largest improvement is in the reduction in the cost of failure.

## 4.5  Quality circles

Quality circles originated in Japan. They provide a method of refining the local and project working conditions and improving productivity. They are based on two factors:

- ❑ the staff can often see problems which are not evident to their managers;
- ❑ the best people to fix a problem are those who stand to benefit from its solution.

There are only two ground rules which must be adhered to if a quality circle is to be successful. First, it must be implemented totally within its own resources, usually at zero cost. Second, tackle one problem at a time, usually the one that yields the highest benefit.

The way quality circles work is to set up a small but representative team from one department. In the case of a project-specific quality circle, the project manager would organise the process. Initially, a brain-storming session would be held to determine any attainable improvements in a project activity. Secondly, the project activity is discussed with other members of the project team. Thirdly, a number of good ideas are identified and proposals made on how the problem may be solved.

Finally, the least expensive solution is implemented within the project as a whole.

In most projects, quality systems are developed as the project proceeds. Many predetermined ideas are often unsuitable for solving particular problems later in the project life-cycle. To ensure that problems are addressed and catered for as they occur, the project manager can utilise the experience of the project team in the quality circle. There can be difficulties in administering this scheme; for example, if all the key personnel on a small project are needed to form an effective quality circle it could create problems in actually organising work on the project. Usually the problems can be overcome, but it should be recognised that a significant commitment in terms of time and resources is required if quality circles are to operate efficiently.

## 4.6  Quality plans

A quality plan is a document setting out the specific quality practices, resources and activities relevant to a particular process, service or project. BS 5750, Part 02, advises that quality plans should define:

❑  the quality objects to be attained;
❑  the specific allocation of responsibility and authority during the different phases of the project;
❑  the specific procedures, methods and work instructions to be applied;
❑  suitable testing, inspection, examination and audit programmes at appropriate stages;
❑  a method for making changes and modifications to a quality plan as projects proceed;
❑  other measures necessary to meet objectives.

To be of value, the first issue of a quality plan must be made before the commencement of work on site. It is also essential that it should be a document which will be read, valued and used by those in control of the work. Their preparation should be started at a very early stage as part of the normal routine of project planning.

## 4.7  Total quality management (TQM)

The steps to total quality management (TQM) ar

❑  understanding quality
❑  commitment to quality

❑  policy on quality
❑  organisation for quality
❑  measurement of the costs of quality
❑  planning for quality
❑  design for quality
❑  systems for quality
❑  capability for quality
❑  control of quality
❑  teamwork for quality
❑  training for quality
❑  implementing TQM

These can be contained in four main elements, management commitment, teamwork, techniques and the quality system, as shown in Figure 4.3. The system is how the other elements are coordinated and enabled to interact.

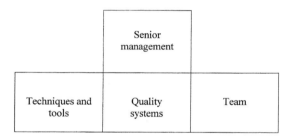

**Figure 4.3** Total quality management.

The main ingredients of TQM are:

❑  quality in meeting the project's requirements;
❑  quality not being an alternative to productivity, but a means of achieving it;
❑  every activity of a project contributing to the total quality of a project;
❑  TQM is a way of achieving project success;
❑  managing quality involves systems, techniques and individuals;
❑  TQM is a way of managing a project.

The advantages of TQM in a project are:

❑  quality in meeting the project specification saves money;
❑  it alleviates quality costs resulting from poor quality;

- costs are reduced by preventing poor quality;
- capacity is increased by improving quality;
- the system provides a basis for teamwork and techniques to interact;
- the system recognises the need to balance risk, benefit and cost;
- the system allows for changes in the project;
- the system records all activities in the project.

Many projects are unique and require the development of quality systems specific to that project. In many cases, existing quality systems can be used as the basis of a project-specific system with the addition of control documentation covering activities not previously encountered. Many project-specific systems are developed as a project proceeds, and thus require continuous updating.

A quality system is itself a management system which can be used by a project manager to ensure that the project meets the quality required through an efficient management control system. When developing a quality system for a project, care should be taken to ensure that:

- the system is appropriate for the project;
- documentation is kept to a minimum;
- the system can be implemented easily and quickly;
- quality plans are succinct and helpful;
- records can be used to help solve disputes;
- reporting procedures and responsibilities are understood by all members of the project team.

## 4.8  Business process re-engineering (BPR) and TQM

TQM was adopted by many organisations during the 1980s, and can be described as continuous process improvement. In the 1990s, many of these organisations began experimenting with more radical change approaches by introducing business process re-engineering. TQM and BPR share many characteristics, such as customer focus, process orientation and commitment to improved performance. However, there are also a number of differences between the two processes, as illustrated below.

- Goal improvement.
  - ■ BPR: radical, sometimes ten-fold levels of improvement in time, cost and quality of process.
  - ■ TQM: around 10% improvement in any given year.

- ❑ Starting point.
  - ■ BPR: new start with a clean sheet of paper.
  - ■ TQM: starting from the current state of the process.
- ❑ Control measurement.
  - ■ BPR: attempts to identify the technological or organisational process factors that will maximise variation and create fruitful change.
  - ■ TQM: stresses the rigour of statistical process control to minimise unexplained variation in each process.
- ❑ Both BPR and TQM are process-orientated.
  - ■ BPR: the focus is on cross-functional processes.
  - ■ TQM: the focus is on functional processes.
- ❑ Rigorous measurement of process performance.
  - ■ BPR: cost of process.
  - ■ TQM: cost of quality.
- ❑ Investment of time before significant results.
  - ■ BPR: time for training and culture change.
  - ■ TQM: time to establish information systems and an organisational structure.

**Table 4.2** Fundamental differences between TQM and BPR.

| Factors | TQM | BPR |
| --- | --- | --- |
| Nature of implementation | Bottom-up | Top-down |
| Main control measure | Cost of quality | Cost of process |
| Type of change | Evolutionary: a better way of doing a job | Revolutionary: a new way of doing business |
|  | Culture change | Culture/structural change |
| Process focus | Functional | Cross-functional |
| Method | Adds value to the existing process | Redesigns the business process |
| Improvement | Incremental (continuous) | Dramatical |
| Risk | Moderate | High |
| Scope of change | Organisation processes | Core business |
| Role of technology | Statistical control | IT enabler |

An on-going polemic regarding the implementation of TQM and BPR is: which comes first? Many project organisations re-engineer processes prior to the introduction of a quality system to ensure that the latest IT enablers are incorporated in the system. If a quality system is developed without sufficient investigation of the processes involved, then it is likely that the system will need major development as the project proceeds.

## Further reading

Ashford, J.L. (1989) *The Management of Quality in Construction*, E & F.N. Spon, London.

BS5750 (1987) *Quality Systems*, British Standard Institution, London.

BS6143 (1992) *Guide to the Economics of Quality*, British Standard Institution, London.

Coulson-Thomas, C.J. (Editor) (1994) *Business Process Re-engineering: Myth and Reality*, Kogan Page, New York.

Hammer, M. and Champy, J. (1993) *Re-engineeering the Corporation: A Manifesto for Business Revolution*, Nicholas Brealey, London.

HM Treasury, Central Unit on Procurement Guidance Note 46, *Quality Assurance*, HMSO, London.

ISO9000 (1987) *Quality Systems*, International Organization for Standardization, Geneva.

ISO10006 (1997) *Quality in Project Management*, International Organization for Standardization, Geneva.

Oakland, J.S. (1989) *Total Quality Management*, Heinemann Professional, Oxford.

# Chapter 5

# Environmental Management

In the past, the main tools for project appraisal were cost–benefit analysis and cost-effectiveness analysis; both logical and quantitative methods for identifying whether or not a project was worth implementing or not. Many publicly funded projects, such as water supply systems, were assumed to have such overwhelming benefits that only the costs were assessed in order to determine which of the various alternative methods of achieving the project objectives was the most cost-effective. It is now recognised that all projects will result in some unquantifiable costs and benefits, and therefore today these appraisal methods are supplemented by environmental impact analysis and assessment. By its nature, environmental impact assessment (EIA) requires a much more qualitative approach to project appraisal than cost–benefit analysis, although this is gradually changing as new methods for valuing environmental impacts emerge and develop.

This chapter covers the main elements of EIA as it is currently used for the appraisal of projects in many countries, and introduces the basic elements of an environmental management system (EMS). Developments in the evaluation and 'monetizing' of environmental impacts are also reviewed and discussed.

## 5.1 Environmental impact

Before discussing environmental impact, it is first necessary to define what is meant by the environment, as it can mean a variety of things to different people – from the light and heat inside a building, to the outdoor environment of natural woodlands, moors, rivers and seas, to the ozone layer surrounding the planet. In the context of the environmental impact of projects, and the discipline of environmental impact assessment, the term environment is taken in its widest context to include all the physical, chemical, biological and socio-economic factors that influence individuals or communities. It not only includes the air, water and land, and all living species of plants, animals, birds, insects, and microorganisms, but also man-made artefacts and structures, and factors of importance to the social, cultural and economic aspects of human existence.

In this context, all projects have an effect or impact on the environment – indeed it could be argued that unless they did there would be no point in implementing projects at all. Some of these may be seen as positive impacts (benefits), while others will have a detrimental effect (costs). The purpose of environmental impact assessment is to evaluate these positive and negative effects as objectively as possible, and to present the information in a manner which is accessible to decision makers so that it can be used in conjunction with other appraisal tools, such as cost–benefit analysis.

Engineering projects may have an impact on the full range of environmental features, including air, water, land, ecology, sound, human aspects, economics and natural resources. Many of these impacts can be measured in terms of changes to specific quality parameters such as the concentration of particulates or hydrocarbons in the air, or the dissolved oxygen concentration in water. Other impacts, such as the aesthetic qualities of a landscape or a structure, or the importance of preserving an historic building, are not so easily quantified.

Some of the impacts will be directly attributable to the project – such as noise from an airport or road, the visual impact of a structure, or pollution of the air or water by a factory. These are referred to as *direct* or *primary* impacts. Other impacts may arise indirectly through the use of the materials and resources required for the project. Examples include the pollution of the air from the manufacture of the cement used for a project, or the impact of quarrying for the raw materials needed for the project. These impacts may affect areas remote from the project itself, and are termed the *indirect* or *secondary* impacts. Lower-order impacts can also be identified, such as the impacts on the environment caused by the manufacture of equipment used for quarrying or cement manufacture.

The range of environmental impacts arising from a project will continue throughout the operating life of the project, and can therefore be regarded as permanent or long-term impacts. The pollution of the air caused by the operation of a thermal power station is thus an example of a long-term impact. Short-term or temporary impacts, on the other hand, are those arising from the planning, design and construction phases of the project. Typically, these might include the noise and dust generated by the construction process itself, or temporary changes to water-table levels, for example.

Environmental impacts may therefore be temporary and direct, temporary and secondary, permanent and direct, or permanent and secondary, and, as discussed later in this chapter, it is necessary to differentiate between these categories of impact within an environmental impact statement.

## 5.2   Environmental impact assessment (EIA)

Environmental impact assessment (EIA) is a logical method of examining the actions of people, and the effects of projects and policies on the environment, in order to help ensure the long-term viability of the earth as an habitable planet. EIA aims to identify and classify project impacts, and predict their effects on the natural environment and on human health and well-being. EIA also seeks to analyse and interpret this information and communicate it succinctly and clearly in the form of an *environmental impact statement* (EIS), which can be used by decision makers in the appraisal of projects.

The natural environment is not in a completely steady state; changes occur naturally over time, some extremely slowly, but others at a much faster rate. Therefore any study of the impacts of a project on the environment must be seen in the context of what would have happened if the project had not been implemented. The environmental impact is thus any environmental change that occurs over a specified period, and within a defined area, resulting from a particular action, compared with the situation as it would have been if the action had not been undertaken.

The report of all environmental impacts that are predicted to arise from a particular project is normally termed an EIS, and is now often required in order to obtain sanction for a project. In many countries, this procedure has now been incorporated in the legal processes of obtaining project and planning approval, and in other situations it has been used voluntarily as part of general project preparation and evaluation.

EIA is a process rather than an activity occurring at one moment during the project cycle. The various steps in the process are shown in Figure 5.1. The process starts with *screening*, which should be done in the early stages of a project as it is concerned with determining whether or not a detailed EIA is required or necessary.

*Scoping* is the second step of the EIA process and is essentially a priority-setting activity. It may be seen as a more specific form of screening, aimed at establishing the main features and scope of the subsequent environmental studies and analysis. The results of the scoping exercise provide the basis and guidelines for the next steps in the process – the base line study and impact assessment itself.

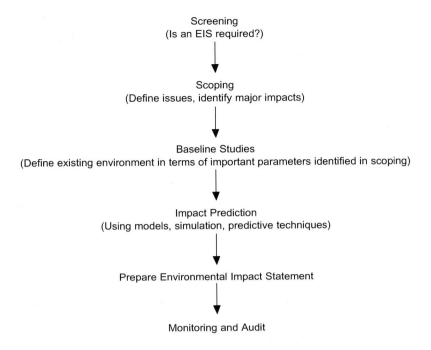

**Figure 5.1** Environmental impact assessment process.

Scoping sets the requirements for the *base-line study*, which is the collection of the background information of the ecosystem and the socio-economic setting of a proposed development project. The activities involve the collection of existing information and the acquisition of new data through field examination and the collection of samples. This step can be the most expensive step of the EIA because it needs a large number of expert people to carry out field surveys and analysis.

The base-line study and the project proposals are then used as the basis for predicting how environmental parameters will change during both the construction and operational phases of the project. Environmental and social scientists will be required for this step, as often sophisticated modelling and simulation techniques must be used.

The predictions and the base-line study are then used to prepare the report – the EIS – which is used in the appraisal and approval processes. Further details of the structure of the report and the more common methods of presenting EIA information are given in subsequent sections of this chapter.

Because of the uncertainty associated with environmental impact predictions, it is important that major environmental parameters are monitored throughout the implementation of the project. This can provide valuable information for future EIAs, and generally improve the accuracy of forecasting models and methods. The process of comparing the impacts predicted in an EIA with those that actually occur after the implementation of the project is referred to as *auditing*.

## 5.3 Screening

Screening is the first step of the EIA process: it is the selection of those projects which require an EIA. For example, the construction of a simple building rarely requires an EIA, whereas the construction of a large coal-burning power station would need a full-scale EIA to be done and an EIS to be prepared, which would include the results of the scientific and technical studies required, supporting documentation, calculations and background information.

The extent to which the EIA is needed for a project is also defined in this step, as the impact of a project depends not only on its own specific characteristics, but also on the nature of the environment in which it is set. Therefore the same type of project can have different impacts, or a different intensity of some impacts, in different settings. So, for any particular project, a full-scale environmental impact assessment may be required for one site but not for another.

The requirements for an EIA also largely depend on the legislative policy of the country in which the project is to be carried out. Through the screening process, only those projects that require an EIA are selected for study. By doing this, unnecessary expenditure and delay is avoided.

The screening process will depend on whether or not appropriate legislation exists, and where it does the laws will often differentiate between those types of project that need an EIA and those that do not. The legislation may also define the extent of the study that will be needed for a particular type of project. Where there is no appropriate law relating to EIA, or if the types of project requiring EIA are not specifically defined, the project manager will have to undertake a preliminary study to ascertain the most significant impacts of the project on the environment. On the basis of the results of this study, and after studying the environmental law and assessing the political situation of the country, the project manager will have to decide whether a full-scale EIA is needed.

## 5.4 Environmental legislation

The concept, and eventually the practice, of EIA evolved, developed and was incorporated into legislation in the USA in the 1970s. The idea was subsequently introduced into other countries such as Canada, Australia and Japan, but it was not until 1985 that EIA was fully accepted within the European Community. At that time, a European Community Council Directive 'on the assessment of the effects of certain public and private projects on the environment' was issued. This required all member states to implement the recommendations set out in that Directive by July 1988.

The Directive called for governments of member states to enact legislation that would make EIA mandatory for certain categories of project, and recommended for other types of project. The projects for which an EIA is *mandatory* include:

- crude oil refineries
- gasification and liquefaction of coal and shales
- thermal power stations
- radioactive waste storage and disposal facilities
- integrated steel and cast-iron melting works
- asbestos extraction and processing works
- production of asbestos products
- integrated chemical installations
- motorways and express roads
- long-distance railways
- airports with runways longer than 2100 m
- trading ports and inland waterways, and ports for traffic over 1350 tonnes
- waste disposal for the incineration, chemical treatment or land fill of toxic and dangerous wastes

The categories of project where an EIA is *recommended* include those listed below.

- *Agriculture:* for example, poultry and pig rearing, salmon breeding, land reclamation, water management, afforestation, restructuring of rural land holdings, use of uncultivated land for intensive agricultural purposes.
- *Extractive industries:* deep drilling, extraction of sand, gravel, shale, salt, phosphates, potash, coal, petroleum, natural gas and ores, and installations for the manufacture of cement.

❑ *Energy industries:* production of electricity, steam and hot water; installations for carrying gas, steam and hot water; overhead cables for electricity; storage of fossil fuels; production and processing of nuclear fuels and radioactive waste; hydroelectric installations.

❑ *Processing of metals:* iron and steel works, non-ferrous and precious metals installations, surface treatment and coating of metals, manufacture and assembly of motor vehicles, shipyards, aircraft manufacture and repair, manufacture of railway equipment.

❑ *Manufacture of glass.*

❑ *Chemical industries:* production of pesticides, pharmaceuticals, paint, varnishes, peroxides and elastomers; storage of chemical and petrochemical products.

❑ *Food industries:* manufacture of oils and fats, packing and canning, dairy product manufacture, brewing, confectionery, slaughter houses, starch manufacturing, sugar factories, fish-meal and fish oil factories.

❑ *Textile, leather, wood and paper industries.*

❑ *Rubber industries.*

❑ *Infrastructure projects:* urban and industrial estate development, cable cars, roads, harbours, airports, flood-relief works, dams, water-storage facilities, tramways, underground railways, oil and gas pipelines, aqueducts, marinas.

❑ *Miscellaneous:* hotels, holiday villages, race and test tracks for cars and motor cycles, waste disposal installations, waste-water treatment plants, sludge disposal, scrap iron storage, manufacture of explosives, engine test beds, manufacture of artificial fibres, knackers' yards.

In the UK, these requirements for EIA have been incorporated into the existing planning legislation, and the developer or initiator of a project now has to supply the relevant environmental impact information in order to obtain sanction for the project. Prior to the enactment of the legislation, planning procedures only addressed land-use considerations, but now issues of pollution control and environmental impact must also be considered.

It is worth noting that environmental impact issues must also be addressed in many overseas projects, either as a result of relevant legislation having been introduced by particular countries or, perhaps more importantly, due to the fact that bilateral funding agencies (such as the UK's Department for International Development (DfID)) and multilateral funding agencies (such as the World Bank and the Asian Development Bank) now have requirements for EIAs built into their loan and grant approval procedures.

## 5.5  Scoping

The US National Environmental Protection Agency (NEPA) Council on Environmental Quality defined scoping as 'an early and open process for determining the scope of issues to be addressed and for identifying the significant issues related to a proposed action'. Scoping is therefore intended to identify the type of data to be collected, the methods and techniques to be used, and the way in which the results of the EIA should be presented.

The question of what is a significant impact is not always an easy one to answer, and requires judgement, tact and an understanding not only of the technical issues, but also of the social, cultural and economic ones. It is likely to involve value judgements based on social criteria (aesthetics, human health and safety, recreation, effect on life styles), economic criteria (the value of resources, the effect on employment, the effect on commerce), or ethical and moral criteria (the effect on other humans, the effect on other forms of life, the effect on future generations).

The base-line study and the impact assessment are expensive and time-consuming, and the determination of the impacts of a proposed project or policy depends on the current state of knowledge relating to any particular aspect of the environment. Very often this depends on the latest developments in the science and technology of environmental monitoring and prediction, as well as the professional experience and judgement of those who carry out the EIA. It is therefore important to ascertain what can reasonably be carried out within the time and budget allowed in order that best use can be made of available resources. This is one of the tasks of the scoping exercise. The other main task is that of determining which are the most important impacts that may occur, and this will vary with the environment surrounding the project. As it is the people living in or near the area of a proposed project who are most likely to have detailed local knowledge and to be most concerned about the local environment, it is extremely important to solicit local public opinion about the project and its perceived impacts on the environment. This is best done at the scoping stage, as local knowledge is a valuable source of data for the base-line studies.

There are two major steps to take in seeking public opinion, the first of which is to identify the population/target groups affected and attempt to obtain their opinions. Local political and environmental concern groups should be consulted at an early stage to avoid unnecessary confrontation at a later stage. In order to carry out this public consultation exercise, a schedule of meetings with the population and target groups affected needs to be arranged. Care should be taken to ensure that all affected and interested groups are included in the programme. At these consultation meetings, the objectives, possible impacts and activities associated with the project should be explained, and everyone should be encouraged to express their opinions on the project and its possible impacts, and to suggest mitigation measures. Sufficient time should be allowed for these meetings, and all suggestions and comments should be recorded in meeting minutes and made public. Major impacts and areas of concern may be identified from this public consultation exercise.

This opinion and data collection procedure is quite a time- and money-consuming activity, and requires patience and diplomacy. In many cases, the EIA has to be completed in a very short period of time, so that meetings with the people affected and other interested groups can become very difficult, if not impossible. In such cases, sample questionnaire surveys can be carried out and, using statistical methods, probable results may be formulated. Where appropriate, it may be possible to carry out some of these surveys by telephone.

A method used in Canada to ensure the public's involvement is to use a panel of four to six experts, who are selected to examine the environmental and related implications of a particular project. The panel is then responsible for issuing guidelines for preparing the EIS and reviewing it after completion. The panel will collect public views through written comment, workshops or public meetings before completing the guidelines for the EIS.

The advantage of scoping is that it helps to obtain advance agreement on the important issues to be considered in the following activities of the EIA, and helps to use scarce resources efficiently in order to analyse the impacts while preventing unnecessary investigation into the issues/impacts which have little importance. Like good early project planning, scoping can save enormous expense and time during the later and most costly steps of EIA. Good scoping requires planning, using competent staff and providing adequate resources.

The outcome of a scoping exercise should be a list of priority concerns and guidelines for the preparation of the EIS, including a structure for the base-line study.

## 5.6  Base-line study

The scoping exercise should provide information about which data are required and which data are unnecessary. If this information is not provided by the scoping team, a great deal of time and money can be wasted in gathering irrelevant and unnecessary data. Unfortunately there are a large number of examples of EIA base-line studies which have included a lot of fairly irrelevant material simply because those preparing them have included all the data that were available, even if they were unnecessary. The base-line study should concentrate on gathering the important data as identified in the scoping exercise. Some of this will be available from previous studies or other sources, but it may be necessary to conduct extensive monitoring exercises to gather other necessary data. The thing to avoid is the costly gathering of data that are of little or no importance.

The base-line study should result in a description of the existing environment, and it should be remembered that this is not static; changes will occur even without the proposed project. The information should be relevant to the impacts discussed, and one should be selective in the information used to describe the base-line. For example, there is no need to spend a great deal of time and energy on data collection and descriptions of present and future noise levels if the project is very unlikely to generate any additional noise. The base-line description should be confined to those impacts and parameters that matter. Too often, the focus of the base-line survey has been on what is available rather than on what is needed.

The outcome of the base-line study is a form of environmental inventory based on primary environmental data collection where appropriate, the opinions of pre-selected individuals and groups, and other sources such as the monitoring and evaluation data from similar completed projects.

As the impacts of the proposed project on the environment are forecast on the basis of the data collected during the base-line study, it is important that the study is carried out accurately and thoroughly.

## 5.7  Impact prediction

For each alternative, the consequences or impacts should be predicted using the most appropriate methods, which may include predictive equations, modelling and simulation techniques. The predictions should include, and differentiate between, primary, secondary and short- and long-term effects. There should be an attempt to show the effect of all project activities on a comprehensive range of environmental parameters, which should have been identified during the scoping stage. These parameters may be grouped into various categories such as biological, physical, social, economic and cultural, and wherever possible quantitative predictions should be made.

Decision makers need to address the question of the accuracy of these predictions – imprecise predictions can generally be counted on to be fairly accurate, whereas there is often a degree of uncertainty associated with precise forecasts. For example, if a power station is being considered, an imprecise prediction such as 'the levels of particulate concentration in the air in the immediate vicinity will increase' is likely to be very accurate. However, making a precise prediction of the value of the particulate concentration will almost certainly suffer from inaccuracies. Some assessment of the accuracy of predictions could be included.

## 5.8 Environmental impact statement

The EIS itself is a detailed written report on the impact of the project on the environment and should contain the following points.

❑ The need for the project: a description of the aims and objectives of the project, explaining clearly why the project is required.
❑ The base-line study report: a comprehensive description of the existing environment, indicating the levels of important environmental parameters, quantitatively wherever possible.
❑ A list and description of all reasonable and possible project alternatives, including the 'do nothing' option. Even options that might be considered 'non-viable' should be mentioned, and reasons given for their rejection. For each of the main alternatives, the following details should be included.
  ■ A clear description of the project during construction and operation, giving details of the use of land, materials and energy, and estimates of expected levels of pollution and emissions.
  ■ A clear estimate of the environmental consequences of each alternative, including predictions of the effects of all the significant impacts on human health and welfare, water, land, air, flora, fauna, climate, landscape and cultural heritage. This section provides the scientific and technical evidence on which comparisons between the alternatives can be made. The report should clearly draw the reader's attention to any serious adverse impacts that cannot be avoided or mitigated, as well as any irreversible environmental consequences and irretrievable use of natural resources.
  ■ The severity of each impact and a clear description of the methods used for measuring and predicting it.

- ❏ The environmental consequences of each alternative should be compared, possibly in tabular or other easily readable form, as discussed later in this chapter. Many consider this to be the most important section of an EIS, and it may often be the first section to be referred to, as it provides an easily understood comparison of all alternatives and their main environmental consequences.
- ❏ The statement conclusions should indicate the preferred option and describe any mitigation measures that may be required to minimise the environmental impacts of this option.
- ❏ A non-technical summary should be included.
- ❏ The report should include an index and any necessary appendices for detailed calculations and data analyses.

## 5.9  Presenting EIA information

One of the difficult tasks in preparing an EIS is that of presenting the information in a way that is comprehensive, yet readily understood by decision makers. This is particularly true of comparisons of the severity of the impacts of alternative proposals. A number of commonly accepted ways of doing this have emerged, and because of their critical importance within the EIS, they are often referred to as EIA 'methods'. These include a number of different forms of check-lists, overlay mapping techniques, networks, multi-attribute utility theory and matrix methods.

### Check-lists

Check-lists are one of the oldest types of EIA method and are still in use in many different forms. Check-lists usually consist of lists (prepared by experts) of environmental features that may be affected by the project activities.

Sometimes a list of project actions that may cause impacts is also incorporated. Check-lists may be a simple list of items, or a more complex list of variations that incorporate the weighting of impacts. Five types of check-list are commonly in use.

- ❏ *Simple lists.* A list of environmental factors and development actions, with no guidance on the assessment of the impacts of the project on these factors. The principal use of this method is to focus attention, and to ensure that a particular factor is not omitted in the EIS.
- ❏ *Descriptive check-lists.* This type of check-list gives some guidance on the assessment of impacts, although it does not attempt to determine the relative importance of those impacts.

❑ *Scaling check-lists.* These consist of a list of environmental elements or resources, accompanied by criteria for expressing their relative value. For each alternative being assessed, the appropriate criteria are selected and any impact that exceeds the defined limit is considered to be significant and should be highlighted for the decision makers' attention.

❑ *Scaling–weighting check-lists.* A panel of experts decides on the weight to be assigned to each environmental parameter. The idea is to assign a weighted score or scale to each possible impact to enable one impact to be compared with another, and to provide an overall environmental impact score for each alternative.

❑ *Questionnaire check-lists.* This type of check-list consists of a series of direct linked questions which are posed to a variety of professionals, who are asked to respond yes, no or unknown to each of the questions.

### Overlay mapping

This procedure consists of producing a set of transparent maps showing the environmental characteristics (e.g. physical, social, ecological, aesthetic) of the proposed project area. A composite characterization of the regional environment is produced by overlaying these maps, and different intensities of impact are indicated on the maps by different intensities of shading.

This method is suitable for the selection of routes for new highways or electrical transmission lines, for example.

One of the main difficulties of using this method is that of superimposing all the maps in a comprehensive way, especially when the number of maps is large. However, by using computers, these problems can be overcome, and a number of computer packages have been developed for this purpose.

### Networks and systems diagrams

These methods were developed as a way of differentiating secondary, tertiary and higher-order impacts from the initial impact. The methodology starts from a list of project activities, and then establishes cause–condition–effect relationships. In the case of systems diagrams, these effects are quantified in terms of energy flows. It attempts to recognize a series of impacts that may be triggered by the project. By defining a set of possible networks, the user can identify impacts by selecting the appropriate project actions.

The advantage of this approach is that it can show the inter-dependency of parameters, and the effects of changes in one parameter on other parameters. The limitations are that it is only really suitable for the assessment of ecological impacts, and it is expensive, time-consuming and requires periodic updating. Systems diagrams depend on a knowledge of the ecological relationships in terms of energy flow, which is often very difficult to characterise.

### Multi-attribute utility theory

Utility theory in EIA has been applied most often to site selection, especially for projects such as major power stations and waste disposal facilities. Alternative projects have different environmental impacts and exhibit different levels (intensity) of the same impact. These methods provide a logical basis for comparing the impacts of alternatives to aid in decision making. The methodology follows the steps described below.

(1)   Determine the environmental parameters that may be affected and can be measured. For each parameter, use appropriate prediction methods to estimate the value of each parameter after the proposed project has been implemented.

(2)   Establish the desirability or otherwise of different levels of each parameter, and formulate the utility function through a systematic comparison of those different levels. The utility $U_i(x_i)$ of each parameter $x_i$ is measured on a scale of 0–1, where 0 is the lowest utility (i.e. the worst possible situation) and 1 is the highest utility (the best possible situation). This step is highly subjective and relies on the values, knowledge and experience of 'experts'.

(3)   Determine a scaling value for each of the environmental parameters which will reflect the relative importance as perceived by decision makers. This is denoted by $k_i$. Again the scaling factors are highly subjective and value-laden.

(4)   The steps described above have to be carried out for each of the alternatives separately, and the total utility, or a composite *environmental quality index* (EQI), for each of the alternatives is then calculated, as shown below.

$$\text{EQI} = U(x) = \sum_{i=1}^{n} k_i U_i(x_i) \tag{5.1}$$

where   $k_i$ is the scaling factor of parameter $x_i$, $U_i(x_i)$ is the utility function of parameter $x_i$ and $U(x)$ is a multi-attribute utility function.

(5)  Determine which alternative has scored the highest EQI value. This will be the least environmentally damaging and the best of the alternatives.

One of the advantages of this method is that the concepts of probability and sensitivity analysis can be incorporated into it, and by using computers a number of 'what if' scenarios can be examined. The fact that a single number, the EQI, is produced is seen by some as an advantage, because it is easy for the decision maker to compare one alternative with another. Others see this as a major drawback, because it produces what appears to be an objective numerical comparison, but which is in fact based on hidden subjective assessments of the utility factors and the scaling factors. Another criticism of the method is that it militates against public understanding and involvement by being unnecessarily technocratic and complex.

Finally, a further limitation of this methodology is that it assumes that the environmental parameters are independent of each other, which is not the case, and that they are fully dependent on the probability assumptions.

## Matrices

### Simple interaction matrix

This form of matrix is simply a two-dimensional chart showing a checklist of project activities on one axis and a checklist of environmental parameters on the other axis. Those activities of the project that are judged by experts to have a probable impact on any component of the environment are identified by placing an X in the corresponding intersecting cell. Matrix methods were originally proposed and developed by Leopold et al. (1971) of the US Geological Survey, to identify the impacts on the environment of almost any type of construction project. Their matrix, now known as the Leopold matrix, consists of 100 specified actions of the project along the horizontal axis and 88 environmental parameters along the vertical axis.

A simple two-dimensional matrix is, in effect, a combination of two check-lists. It incorporates a check-list of project activities on one axis and a check-list of environmental characteristics which may be affected by the actions of project on the other axis. From the matrix, the cause–effect relationships of actions and impacts can easily be identified by marking the relevant cells. Various modifications have been made to Leopold's original lists of activities and environmental characteristics to suit particular situations.

### Quantified and graded matrix

The original Leopold matrix is slightly more complex in that a number, ranging from 1 to 10, is used to express the magnitude and importance of the impacts in each cell. Thus, a grading system is used in place of the simple X.

Modifications of these matrices have been going on over time with the advancement of EIA. In 1980, another grading system was suggested which incorporated the relative weights of each development activity. This method can identify major activities and areas that need more attention.

The advantages of a matrix are that it can rapidly identify the cause–effect relationships between impacts and project activities, and it can express the magnitude and importance in both qualitative and quantitative forms. Another major advantage of the matrix method is that it can communicate the results to the decision makers as a summary of the EIA process. It can be used for any type of project, and can be modified as required.

## 5.10  Monitoring and auditing of environmental impacts

### Monitoring

Environmental impact assessment is largely concerned with making predictions about the effects projects will have on the environment. Unless the actual impacts are measured and monitored during and after project implementation, and are compared with those predicted, it is not possible to assess the accuracy of predictions, or to develop improved methods of prediction. The monitoring of environmental impacts is thus an essential component of the EIA process.

The aim of monitoring is first to detect whether an impact has occurred or not, and if so to determine its magnitude or severity, and second to establish whether or not it is actually the result of the project, and is not caused by some other factor or natural cause. In order to do this, it is desirable to identify 'control' or 'reference' sites, which should be monitored as well as the project site. The reference site should be as similar as possible to the project site, except that the predicted project impacts are unlikely to occur. To be effective, monitoring should begin during the base-line study, and continue throughout the construction and operational phases of the project.

During the scoping stage, a framework for monitoring should be established. This should outline which parameters to monitor, the required frequency of sampling, the magnitude of change of each parameter that is statistically significant, and the probability levels of changes occurring naturally. The establishment of this framework is perhaps more important than the actual details relating to data measurement and collection, as there is little point in spending a great deal of time and money in gathering irrelevant or insignificant data. The monitoring process can be very expensive.

Whilst the main objective of monitoring is to assess the validity and accuracy of predictions (auditing), it is worth noting that monitoring data can provide an early warning of possibly harmful effects in time to initiate mitigating measures and thus minimise adverse impacts. Monitoring improves general knowledge about the environmental effects of different types of projects on a variety of environment parameters. As more monitoring exercises are carried out, the data available will increase and become more reliable, thus improving future EIA studies.

*Auditing*

Auditing is the process of comparing the predicted impacts with those that actually occur during and after project implementation. Auditing thus requires accurate and well-planned monitoring, as explained above.

Auditing and monitoring are steps in the EIA process that are often disregarded or not given sufficient attention. One of the reasons for this is that EIA is seen by many as a hurdle in the approval and sanction phase of a project, and once this has been granted, the environmental effects are given much less attention. If EIA is to be used as a method of reducing environmental impacts and/or mitigating against them, the auditing process must be given a much higher priority than is currently the case. Auditing not only helps to provide information on the accuracy of environmental impact predictions, but also highlights the best practice for the preparation of future EISs.

Audit studies of EIAs produced in the past have shown considerable variation and inaccuracies in environmental parameter predictions, and a large number of impacts studied could not be audited for a variety of reasons. These included inappropriate forms of prediction, the fact that design changes had been made after the EIS had been completed, and inadequate or non-existent monitoring. In some cases less than 10% of the impacts evaluated in an EIA could be audited, and of these, less than 50% were shown to be accurate. There is therefore great room for improvement in the preparation and presentation of EISs, and effective monitoring and auditing are absolutely essential if these improvements are to be made.

## 5.11  Environmental economics

Decision makers are often faced with two documents: the mainly qualitative report on environmental effects (the EIS), and the 'objective' numerical statement of costs and benefits (the cost–benefit appraisal). Both are concerned with costs and benefits, but they are written and presented in different terms and forms. For this reason, a relatively new science of *environmental economics* has developed, which helps to coordinate these two appraisal tools and perhaps eventually to combine them.

### Cost–benefit analysis (CBA)

In its simplest form, the principles of CBA can be expressed by the following equation:

$$3(B_t - C_t)(1 + r)^{-t} > 0 \qquad (5.2)$$

where $B_t$ is the benefits accruing at time $t$, $C_t$ is the costs incurred at time $t$, $r$ is the discount rate and $t$ is the time.

However, only those costs and benefits that can easily be quantified are normally included, and the environmental and social costs and benefits are ignored. There are two questions to address, the first being whether or not there are indeed any real costs and benefits that can be ascribed to the environment, and the second being how they can be valued in the same terms as other project costs and benefits: i.e. in money terms.

If both these questions can be answered satisfactorily, the CBA equation could be rewritten as

$$3(B_t - C_t \pm E_t)(1 + r)^{-t} > 0 \qquad (5.3)$$

where $E_t$ is the environmental change (cost or benefit) at time $t$.

To address the first question, consider the situation where a community relies on catching fish from an unpolluted river for its income. If a factory sited upstream then discharges waste directly into the river, the likelihood is that the fish population in the river will be affected and, as a direct consequence, the livelihood and welfare of the fishing community downstream will decline. In this instance, there is clearly a cost associated with the pollution of the environment: i.e. the loss of income suffered by the community.

This simple example illustrates that environmental 'goods' are not free. Whenever they are used, either as a resource or as a waste depository, they do incur some form of cost, whether that be in terms of loss of income, or a reduction in health or welfare, as a consequence of air pollution, high noise levels, polluted land or water, or in a number of other ways. The only reason that these costs are not always evaluated is because there is no market for environmental goods – we cannot buy clean air or quiet surroundings, at least not directly. Nevertheless, these things do have value to people, and they increase or decrease their welfare in much the same way as other commodities, such as a new car, a washing machine, or the provision of electricity, affect the quality of people's lives.

But how can environmental costs and benefits be measured in money terms; how can they be 'monetized'? Over the last few years, environmental economists have been developing new approaches to valuing the environment, and many of these techniques can now be used to incorporate environmental costs and benefits in traditional cost–benefit analyses.

## Environmental values

Environmental economists ascribe three distinct forms of value to the environment. The first of these is what is termed *actual use value*, and this refers to those benefits directly derived by people who actually make use of the environment and gain direct benefit, such as farmers, fishermen and hikers. Polluters also derive a benefit from the environment as a consequence of being able freely to dispose of their waste.

The second form of environmental value is *option use value*, which is the value that some people put on the environment in order to preserve it for future use, either by themselves or by future generations. For example, they may believe that a particular area of scenic beauty is worth preserving because, even if they do not use it now, they may want to use it in the future.

In addition to this, some people place a value on the environment for its mere existence, even if they see no probability that they will ever make use of that environmental asset. Some people, for example, place a value on the rainforests, or the preservation of certain endangered species, even though there is no possibility of those people ever seeing them. This is the *existence value* of the environment.

The *total economic value* of an environmental asset is equal to actual use value plus the option use value plus the existence value.

## Methods for valuing the environment

The main methods that have been developed for 'monetizing' environmental costs and benefits are:

- the effect on production methods
- preventative expenditure and replacement cost
- human capital
- hedonic methods
- the travel cost method (TCM)
- contingent valuation

### The effect on production methods

In its simplest form, an example of this method might be the reduction in fish catches due to pollution of a watercourse, as described previously. However, changes in market prices for goods and services must also be considered, as these will influence the net cost or benefit. For example, if increased production saturates the market and prices then fall, the net benefit may be reduced.

The procedure for quantifying the effect on production is a two-stage process. First, the link between the environmental impact and the amount of lost production must be established, and second, the monetary value associated with the changes must be calculated.

It is often very difficult to establish the physical link between the affected environment and the change in output: for example, to establish the relationship between the effect of fertilizers used in an upland catchment area of a lake on diminishing fish stocks in the lake. There may be other causes, and it is often an extremely complex task to disentangle the various factors contributing to a loss in production, particularly when trying to separate man-made from natural effects. Determining the effect requires a 'with' and 'without' scenario to be established. This is particularly difficult, if not impossible, when project actions have already been initiated.

Despite the difficulties mentioned above, this technique is probably the most widely used environmental valuation method.

### Preventative expenditure and replacement cost

This method assesses the value placed on the current environment in terms of what people will spend to preserve it or stop its degradation (defensive expenditure), or to restore the environment after it has been degraded. For example, the cost of additional fertilizers, which are used to compensate for the loss of natural fertility in the soil arising from a particular project, could be used as the environmental cost of the project. Similarly, the cost of installing double glazing in houses affected by noise from a road or airport can be regarded as the minimum cost of this form of environmental pollution arising from the project. In some cases, of course, these costs are real and actually become part of the project cost.

However, even if the preventative measures are not taken, there is a cost to those affected and it should be given consideration in the appraisal of the project.

This is often done by estimating the cost of a shadow or compensating project, which would be required to bring the environment to the same quality as it was before the main project. An example of this might be the drilling of boreholes and the establishment of an irrigation system for people displaced by a dam project from a village that had good natural irrigation to an area where crop irrigation is problematic.

### Human capital
This method considers people as economic units and their income as a return on investment. It focuses on the increased incidence of disease and poor health, and the consequent cost to society in terms of increased health care costs and reduced potential for earnings by the individuals affected. The method suffers from the problem that it is often very difficult to isolate a single cause of productivity loss. For example, poor health may be caused by unclean water, poor domestic hygiene, polluted air or a wide variety of other factors. However, if a direct causal link with a particular environmental pollutant can be proven, the environmental cost of the pollution could be calculated by assessing the loss of earnings of those people affected, and the additional costs of medical care required. Although relevant statistics may be available in developed countries, it is unlikely that the method could be used in many developing countries, owing to the lack of records and inadequate information.

### Hedonic methods
These methods use surrogate prices for environmental 'goods' for which there are no direct markets. They seek to link environmental goods with other commodities that do have a market value. For example, the price of a three-bedroom semi-detached property will vary according to the environment in which it is located. People are prepared to pay more money for a house located in a quiet area with a pleasant view, than they would for essentially the same house located next to a noisy, dirty factory. The housing market can therefore be used as a 'surrogate market' for the environment. There are, of course, many other factors that affect house prices, such as the proximity to schools and transportation routes, and these factors must be isolated and removed, using techniques such as multiple regression analysis, if the method is to prove in any way reliable. The method thus requires large amounts of data, which are then very difficult to analyse. However, it has been used

successfully to evaluate the costs of noise and air pollution on residential environments.

### Travel cost method

This method is particularly appropriate for valuing recreational areas with no or low admission charges. The value of the site is calculated as the cost incurred by the visitors in both travelling expenses and time spent travelling. The further people travel to the site, the greater value they place on it, as they will use more fuel, incur higher fares and spend more hours in transit to and from the site.

Surveys must be carried out at the site in question. Visitors are asked how far they have travelled, how long the journey took, what type of transport they used, whether they visited any other locations, and how often they visit the site. The results of the survey, together with information on the total number of visitors to the site each year, are used to determine an annual total cost incurred by all visitors to the site. It is assumed that this cost must be a minimum value that is placed on the amenity.

This method is mainly used to determine the value of preserving national parks and other sites of scenic, recreational and scientific interest.

### Contingent valuation method

The contingent valuation method is a technique that can be used where actual market data are lacking. The method involves asking people what their preferences are, how much they are willing to pay ('willingness to pay', WTP) for a certain environment benefit, and how much they would need to be paid ('willingness to accept', WTA) in compensation for losing a certain environment benefit.

The technique usually involves questionnaires being sent to the parties concerned or house-to-house interviews. This technique may be used on its own or in conjunction with the other techniques described above.

### The future

EIA is now a recognised and accepted practice, and is used widely in appraising projects, primarily at the sanction stage. In most countries it is now incorporated into the legislation relating to planning and approval, and many international funding agencies now require an EIS as part of their loan or grant approval procedures.

If EIA is to continue to develop and to maintain credibility, it is essential that it is seen not just as a means to obtain a permit to develop, but as an environmental management tool to be used throughout the life of the project: from inception to design, construction, operation and decommissioning. This will only come about if monitoring and auditing are conscientiously carried out, and the results are fed back into the EIA process.

The valuation of environmental impacts is likely to improve as more data are gathered and confidence in the methods grows. It is already possible to incorporate some environmental impacts as costs and benefits in conventional cost–benefit analysis. The sceptics point to inaccuracies associated with the monetary evaluation of environmental impacts, but even the direct costs of design and construction cannot be predicted accurately. The way ahead is to recognise that, as with direct costs, social and environmental costs can only be estimated within certain limits of confidence, and to build these ranges, rather than discrete values, into our cost–benefit appraisal calculations.

## 5.12  Environmental management

Environmental management is not separate from project management; it must be seen by project managers, design engineers and constructors as an integral part of their work at all stages of the project cycle. Although some elements of EIA, such as monitoring and auditing, should be an integral part of the design, construction, operation and decommissioning phases of a project, EIA is still used primarily at the feasibility, appraisal and sanction stages of the project cycle. However, EIA is not the only 'tool' or process that can be used for environmental management, and one that has recently risen to prominence throughout the world, and in a wide range of industries, is the environmental management system, or EMS.

Whereas EIA is an analysis and appraisal tool that enables decision makers to predict the environmental impacts of a project, an EMS is an action tool, the purpose of which is to provide managers and other decision makers with clear guidelines on how to improve an organisation's environmental performance, in every aspect of the work they do (which may include carrying out projects). An EMS is part of an organisation's overall management system, and as such it includes strategic planning, organisational structure and the implementation of an organisation's environmental policy as an integrated part of the organisation's activities. In this sense, an EMS bears a lot more resemblance to a quality management system than to an EIS. The idea of the EMS was first formally recognised and given authority by the publication in 1992 of British Standard *BS 7750. Specification for Environmental*

*Management Systems.* This was superseded first in 1994 by a new version of BS 7750, and again in 1996 by the international standard EN ISO 14001, one of the ISO 14000 family of documents on environmental management.

It should be stressed that whereas EIA is now a legal requirement for certain types of project in most countries, there is no legal requirement for a company or organisation to set up an EMS. However, it is becoming more and more common for major clients to require their contractors to have an EMS and, as a consequence, for contractors to require their sub-contractors and suppliers to have one also. Thus, the spread of the use of the EMS is being driven by the supply chain rather than by legislation.

The ISO 14001 standard requires a company or organisation to establish and maintain a system that enables it to:

❑   establish an environmental policy;
❑   identify the environmental aspects of their operations;
❑   determine the *significant* environmental impacts;
❑   identify relevant laws and regulations;
❑   set environmental objectives and targets;
❑   establish a structure and programme to implement objectives;
❑   facilitate planning, control, monitoring, review and audit;
❑   be capable of adapting to change.

Like a quality management system, an EMS is a management standard; it is not a performance or product standard. The underlying purpose of ISO 14001 is that companies will improve their environmental performance by implementing an EMS, but there are no standards for performance or for the level of improvement. It is a process for managing company activities that impact on the environment. Figure 5.2 illustrates the various steps necessary to set up an EMS for an organisation.

## Environmental policy

The first step in implementing an EMS is to gain a commitment to environmental improvement from top-level management, and this must then be made public by issuing a corporate environmental policy statement. A preliminary review of all aspects of the organisation in the light of strengths, weaknesses, risks and opportunities in relation to environmental issues and effects should then be undertaken. The review should cover all legislative and regulatory requirements and an

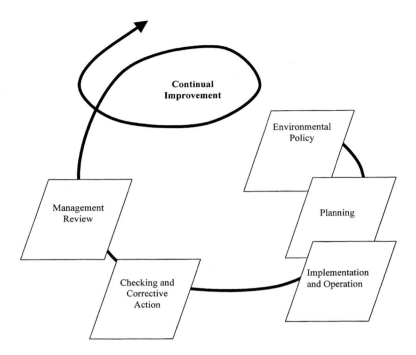

**Figure 5.2** The steps needed to set up an EMS (after ISO 14 001).

assessment of previous non-compliance, together with an evaluation of known significant environmental impacts of the organisation's activities, and an examination of existing environmental management procedures.

This review sets the base-line for the organisation's environmental policy. In order to meet the ISO 14 001 standard, the policy must contain a commitment at least to meet all relevant environmental legislation, and to show a continuing commitment to an improvement in environmental performance over and above that required by the legislation. It must also be made public. These are the essential requirements for meeting the standard. It should be noted that the standard does not require an organisation to meet externally set targets for environmental performance (other than minimum legal requirements); the organisation must set these targets itself.

### Planning

ISO 14 001 requires an organisation to set up and maintain programmes and procedures to identify any environmental effects arising from the support functions (planning, finance, sales, procurement) of the organisation as well as the main activities. These effects may be direct, such as the release of pollutants into the environment, or indirect, which may include the possible misuse of the company's products by others, or the environmental impacts arising from the activities of suppliers and sub-contractors. The plan should also include procedures for identifying and ensuring access to relevant environmental legislation and regulations.

A major element of planning is the setting of objectives and targets. These must be consistent with the organisation's policy, they must be documented, and they need to be developed for, and communicated to, all levels of the organisation. Wherever possible they should quantify the targeted levels of improvement in environmental performance. For example, '*We will reduce hazardous waste emissions by 50% of current levels within the next 12 months*'.

Once objectives have been set, programmes to achieve them must be developed and documented. These programmes will formally assign responsibilities, and specify the means and time frame by which they will be achieved.

### Implementation and operation

In order to implement the programme, the necessary resources must be provided, and these must be fully detailed in the plans. A management representative must be appointed to ensure that the EMS is established, implemented and maintained. In addition, individuals need to be nominated to ensure that the organisation is kept up-to-date with all relevant environmental legislation and for training at all levels within the organisation. To conform to the requirements of the standard, organisations must implement procedures for internal communication between the various levels and functions of the organisation, and for dealing with relevant communications from external parties such as suppliers, sub-contractors, environmental groups, public organisations and the general public.

The documentation for the EMS need not be a separate manual; it can be integrated with other management systems such as quality, and health and safety. The important thing is that it is easily available to all employees and, to this end, it may be in paper or electronic format, as appropriate to any given situation.

Operations and activities, including maintenance, that can have a significant impact on the environment need to be identified and then controlled and monitored to ensure compliance with legislation and the organisation's EMS. The organisation should also identify any significant environmental impacts associated with the goods and services that it uses, and communicate the relevant procedures and requirements to its suppliers and contractors. Methods for preventing or mitigating environmental impacts associated with emergencies should also be developed and documented.

### Checking and corrective action

The characteristics of operations and activities that can have a significant impact on the environment need to be monitored and measured regularly. Records of monitoring and measurement are needed to assess performance, to demonstrate that operational controls are effective, and to show that objectives and targets are being met. The results of any monitoring should then be compared with the legislative and regulatory requirements in order to demonstrate compliance. For each environmentally relevant activity, the organisation should document how verification is to be carried out and the required accuracy of the measurements. Actions to be taken in the case of non-conformance should be detailed in the EMS, together with the reasons for non-conformance, and the procedural changes implemented as a result.

The EMS itself must also be subjected to systematic and regular audit in order to provide assurance that the system is being implemented satisfactorily, and to provide information for the management review. Procedures for audit need to be established to identify the frequency of audits, which aspects of the organisation's activities will be audited, who will be responsible for the auditing, and how the results will be documented and communicated. Although an internal auditor may conduct audits, an external auditor, verified by the appropriate agency, must be used in order to gain formal accreditation to ISO 14001. Whether internal or external auditors are used, they must have the support and authority to obtain the necessary information.

### Management review

The EMS cycle is concluded by the management review, which is conducted by senior management to ensure that the system is operating effectively. This provides the opportunity to make any changes to policies, objectives or the EMS itself, all of which may be required due to changes in legislation, modifications to business operations, advances in technology, results of audits, or for continual improvement.

# Reference

Leopold, L.B., Clark, F.E., Hanshaw, B.B. and Balsley, J.R. (1971) A procedure for evaluating environmental impact, US Geological Survey Circular 645, Department of Interior, USA.

# Further reading

Harrington, H.J. and Knight, A. (1999) *ISO 14000. Implementation: Upgrading your EMS Effectively*, McGraw Hill, New York.

Hughes, D. (1992) *Environmental Law*, 2nd edn, Butterworth, London.

ISO (1996) *ISO 14001. Environmental Management Systems: Specifications and Guidance for Use*, International Organization for Standardization, Geneva.

Roberts, H. and Robinson, G. (1998) *ISO 14001. EMS Implementation Handbook*, Butterworth Heinemann, Oxford.

Turner, R.K., Pearce, D. and Bateman, I. (1994) *Environmental Economics: An Elementary Introduction*, Harvester Wheatsheaf, Hemel Hempstead.

Wathern, P. (Editor) (1988) *Environmental Impact Assessment: Theory and Practice*, Unwin Hyman, London.

Winpenny, J.T. (1991) *Values for the Environment: A Guide to Economic Appraisal*, HMSO, London.

Wood, C. (1995) *Environmental Impact Assessment: A Comparative Review*, Longman, Essex.

# Chapter 6
# Project Finance

In this chapter the sources of finance, methods of project financing, types of loan, loans in a mixture of currencies, financial risks and the appraisal of finance for projects are described.

## 6.1 Funding for projects

Public finance has provided most of the major infrastructure projects procured in the UK and overseas in the last 50 years. Monies raised from taxation have provided all or part of the finance required for projects such as motorways, bridges and tunnels, transport systems, and water and power plants. In many cases, low-interest-rate loans or deferred or subordinated loans were provided by government, in conjunction with loans from the private sector, for major infrastructure projects, resulting in a hybrid, mixed funding finance package. The fact that no money is free, even to governments who borrow from the private sector and bear the costs of collecting monies themselves, has resulted in private finance being considered for a number of major projects normally procured with public funds.

The use of private finance instead of public finance for a particular project is only justified if it provides a more cost-effective solution. The financial plan of a project will often have a greater impact on its success than the physical design or the construction costs. The cost of finance and its associated components are often determined by the method of revenue generation, the type of project and its location.

Often, government support to lenders of project finance, in the form of guarantees, has been sufficient to ensure that projects have been completed that would not have been commercially viable without such support.

## 6.2 Sources of finance

In major projects, finance is often provided by a lender in the form of a commercial bank, a pension fund, an insurance company, an export credit agency or a development bank. Other project finance sources include institutional investors, large corporations, investment banks, niche banks and developers, utility subsidiaries, and vendors and contractors. The more attractive, in terms of potential returns, the project is often increases the number of sources of finance. Figure 6.1 shows a number of sources of funding for projects.

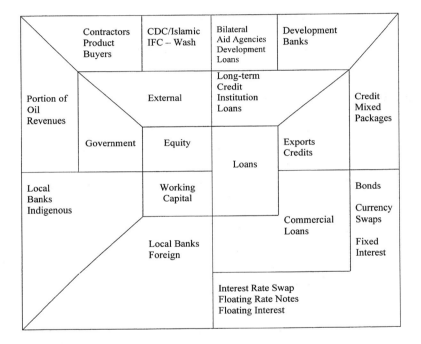

**Figure 6.1** Sources of finance.

In the Channel Tunnel project, for example the debt/equity ratio was initially set at 80/20, with finance being raised from the sources listed below.

❏ Equity
- Banks and contractors
- Founder shareholders
- Private institutions, 1st tranche
- Public investors, 2nd tranche

- ■ Public investors, 3rd tranche
- ■ Public investors, 4th tranche
- ❏ Debt
  - ■ Commercial banks, main facility
  - ■ Commercial banks, standby facility

In order to pay the loan arrangements, revenue is obtained from three sources, i.e. shuttle services, British and French railways, and other associated facilities. The original financing structure of this project involved 210 lending organisations, with 30% of the loans being arranged through Japanese banks.

In some cases where no equity finance is deemed to be necessary, finance will be provided purely on the basis of debt, and is normally arranged by one lender acting as a lead lender for a number of banks or institutions. In many cases, governments or multi-national guarantee agencies may guarantee loans over specific lending periods for specific types of project. Many lender organisations will only lend to projects with reasonably assured cash flows and sufficient government support.

## 6.3  Project finance

Project finance is the term used to describe the financing of a particular legal entity whose cash flows and revenues will be accepted by the lender as a source of funds from which the loan will be repaid. Thus, the project's assets, contracts, economics and cash flows are segregated from its sponsors so that it is of strictly limited recourse, in that the lenders assume some of the risk of the commercial success or failure of the project. In summary, project finance may be defined as

> the financing of a particular economic unit in which the lender is satisfied to look initially to the cash flows and earnings of that unit as the source of funds from which the loan will be repaid and to the assets of the economic unit collateral for the loan. (Nevitt, 1983)

Project finance provides no recourse, and if the project revenues are not sufficient to cover the debt service, the lenders have no claim against the owner beyond the assets of the project; the project, in effect, being self-funding and self-liquidating in terms of financing.

Unlike traditional public sector projects whose capital costs are largely financed by loans raised by government, privately financed projects are financed by a combination of debt and equity capital, and the ratio between these two types of capital varies between projects.

Financing a project is often the most difficult operation, with the largest risks occurring during the construction phase. Often, equity finance is arranged to overcome this problem before revenue is generated by the project. This is similar to the conditions for the original loan companies, and it is often a requirement of a promoter organisation that the parent company should guarantee performance by the sponsor. Similarly, lending organisations must look to guarantees to cover such risks on loans provided to sponsors.

A private finance package is dependent on the type of project, the currencies of the loan, the loan schedule and the possible effect of associated risks. If private finance is to become a major feature in procuring infrastructure projects which are normally undertaken by the public sector, it is important that each project and hence finance package, is considered on an individual basis. Government support must be given to projects that would not be commercially viable in the private sector to ensure that the social and economic benefits of projects are enjoyed by the users.

The basic features of project finance are summarised below.

(1)  Special project vehicle (SPV).
(2)  Non-recourse or limited recourse funding.
(3)  Off-balance-sheet transaction.
(4)  Sound income stream of the project as the predominant basis for financing.
(5)  A variety of financial instruments.
(6)  A variety of participants.
(7)  A variety of risks.
(8)  A variety of contractual arrangements.

## 6.4  Financial instruments

All projects require financing. No project progresses without financial resources. However, the nature and amount of financing required during different phases of the project varies widely. In most projects, the rate of expenditure changes dramatically as the project moves from the appraisal stage, which consumes mainly human expertise and analytical skills, to the design stage, then to the manufacture and construction phase, and finally to the operational phase. Broadly, a project may be said to pass through three major phases:

❑  project appraisal;
❑  project implementation/construction;
❑  project operation.

The precise shape of the cash flow curve for a particular project depends on various factors such as the time taken in setting up the project objectives, obtaining statutory approvals, design finalisation, finalisation of the contracts, finalisation of the financing arrangement, the rate and amount of construction, and the operation speed. The negative cash flow until the project breaks even clearly indicates that a typical project needs financing from outside until it does break even. The shape of the curve also reveals that in the initial phase of the project, relatively less financing is required. As the project moves on to the implementation phase there is a sudden increase in the finance requirement, which peaks at the completion stage. The rate of spending is also depicted by the steepness of the curve. The steeper the curve, the greater the need for finance to be available. Once the project is commissioned and starts to yield revenues, the requirement of financing from outside becomes less and less. Finally, the project starts to generate sufficient resources for its operation and maintenance and also a surplus. However, even after the break-even point, the project may require financing for short periods to meet the mismatch between receipts and payments.

In project financing, it is this future cash flow that becomes the basis for raising resources for investing in the project. It is the job of the project finance team to package this cash flow in such a way that it meets the needs of the project, and at the same time is attractive to the potential agencies and individuals willing to provide resources for investment. In order to achieve this objective effectively, a thorough knowledge of the financial instruments and financial markets in which they trade is essential.

Projects have to raise cash to finance their investment activities. This is normally done by issuing or selling securities. These securities, known as financial instruments, are in the form of a claim on the future cash flow for the project. At the same time, these instruments have a contingent claim on the assets of the project, which acts as a security in the event of future cash flows not materialising as expected. The nature and seniority of the claims on the cash flow and assets of the project vary with the financial instrument used. Merna and Owen (1998) describe financial instruments as 'the tools used by the SPV to raise money to finance the project.' Traditionally, financial instruments were in the form of either debt or equity. Developments in the financial markets and financial innovations have led to the development of various other types of financial instrument which share the characteristics of both debt and equity. These instruments are normally described as 'mezzanine finance'.

*Debt* instruments refer to those securities issued by the project that make it liable to pay a specified amount at a particular time. Debt is senior to all other claims on the project cash flow and assets. Debt instruments include term loans, Euro currency loans, debentures, export credit, supplier's credit and buyer's credit.

Ordinary *Equity* refers to the ownership interest of common stock-holders in the project. On the balance sheet, equity equals total assets less all liabilities. It has the lowest rank and therefore the last claim on the assets and cash flow of the project. Equity is normally in the form of ordinary shares and preference shares.

*Mezzanine finance* occupies an intermediate position between the senior debt and common equity. Mezzanine finance typically takes the form of subordinated debt, junior subordinated debt and preferred stock, or some combination of each. Mezzanine finance instruments include plain vanilla bonds, junk bonds, floating rate bonds, discount bonds, income bonds, Euro bonds, revolving underwriting facility (RUF), note issuance facility (NIF), warrants and convertible bonds. As well as debt, equity and mezzanine finance, a project may also utilise certain other types of instrument such as leasing, venture capital, aid and depository receipts.

## 6.5  Financial engineering

Just as engineers use special tools and instruments to achieve engineering perfection, the financial engineers use specialised financial instruments and tools to improve financial performance. The term financial engineering can be defined as 'the development and creative application of financial technology to solve financial problems and exploit financial opportunities', or 'the use of financial instruments to restructure an existing financial profile into one having more desirable properties'.

Financial engineering techniques are being put to wide applications such as modelling and forecasting financial markets, development of derivative instruments and securities, hedging and financial risk management, asset allocation and investment management, and asset/liability management.

The tools used by financial engineers comprise the new financial instruments created during the facility for the Channel Tunnel project, under which 70% of the project equity had to be spent before any loans could be taken up by the promoter. The 'offer of sale' of equity shares made no mention of shareholders entitlements after the concession is transferred to the UK and French governments in July 2042. This is characteristic of investment equity capital, since unlike other forms of capital market lending, the loan never comes to maturity, i.e. the loan is permanent and is never repaid. However, equity shares are negotiable and may be transacted on the Stock Exchange.

## 6.6  Debt financing contract

The contract between the sponsor and lender can only be deter- mined when the lender has sufficient information to assess the viabi- lity of a project. In construction projects, the lender will look to the project itself as a source of repayment rather than the assets of the project. The key parameters to be considered by lenders include those listed below.

- ❑ The total size of the project: the size of the project will determine the amount of money required and the effort needed to raise the capital, the internal rate of return on the project and the equity.
- ❑ Break-even dates: critical dates when equity investors see a return on their investments.
- ❑ Milestones: significant dates related to the financing of the project.
- ❑ Loan summary: the true cost of each loan, the amount drawn and the year in which drawdowns reach their maximum.

A properly structured financial loan package should achieve the fol- lowing basic objectives.

- ❑ Maximise long term debt: allow the project entity itself to incur the debt, which would not affect the balance sheet of the sponsors' parent company.
- ❑ Maximise fixed-rate financing: the utilisation of long-term export credit facilities or subordinated loans with low interest rates over long terms will reduce project risks.
- ❑ Minimise refinancing risk: cost overruns present additional problems to a project, and therefore standby credit facilities from lenders and additional capital from promoters should be made available.

In order to manage various types of risk, a lender will often prepare a term sheet. This term sheet, based on agreement among the project participants, defines the rights and obligations of lenders, and describes default conditions and remedies, and serves as the bid document for accessing capital markets.

The components of a term sheet determine whether a loan would be sanctioned for a particular project, and provide a mechanism for organisations to evaluate the financial parameters of a bid. These parameters may include a bank loan, index or coupon bonds, drawdown of loans, equity, dividends, preferred shares, interest during construction, interest during operation, lenders fees, interest rate risk (IRR), net present value (NPV), the coverage ratio, the payback period of loans, a standby facility, the working capital and the debt service ratio.

Lenders will need to be satisfied that the project can be seen to be viable for each risk analysis. In assessing the adequacy of projected cash flows, the major criteria may be:

- the debt service ratio (defined as the annual cash flow available for debt service) divided by the debt service;
- the coverage ratio (defined as the NPV of future after-tax cash flows over either the project life or the loan life) divided by the loan balance outstanding.

While the debt/service ratio is used in the evaluation of limited resource financing in all types of industry, the coverage ratio is a well-established criterion for the evaluation of oil and gas projects.

For example, in a project where the coverage ratio is 1:1, the debt will be repaid with no margin of error in the cash-flow projections. Ratios of less than one mean that the debt will not be repaid over the term of the calculation, and ratios in excess of one provide a measure of comfort should variations from the assumptions occur. The term sheet may be used to express such ratios for one or a number of risks analyses, and should the ratio be less than the lender's requirement, then the loan package may be considered commercially non-viable for the risks to be covered.

## 6.7 Types of loans

The loan structure of a finance package is often the most important ingredient in the success of a project. The structure of the loan may be in any of the forms described below.

### Debt: loan and debentures

The conditions of loan finance will depend on the criteria of the lender and the sponsor and the type of project considered. The main feature that will need to be agreed is the loan repayment method.

## Repayment method

A *mortgage* may be set up with repayments calculated so that the total amount paid at each instalment is constant, while capital and interest vary. Initially, this means that the proportion of capital repaid is small, but progressively increases throughout the project. This form of loan is suitable for projects where no revenue is generated until the project is operational.

With *equal instalments of principal*, the amount of principal repaid is constant with each payment, with the total amount paid decreasing over time as it consists of a constant principal repayment plus interest on outstanding principal.

With a *maturity* loan, the principal is repaid at the end of the loan period in one sum. This form of loan is most suitable for projects which generate a large capital sum on completion, and could only be considered for projects where the sponsor intended to sell the facility immediately after commissioning.

Other variations of repayment structures may include a moratorium on capital repayments or interest payments for a period at the start of the loan, a bullet payment, or a sunset payment at the end of the loan period. This structure is useful for projects which do not begin to generate revenue until fully operational.

## Interest rate

The interest rate will often be governed by the money market rates, which vary considerably. Interest rates may be fixed for the period of the loan, or expressed as a percentage of the standard base rate. In some cases, the sponsor may negotiate a floating rate where an upper and lower limit are set on the interest rate. These floating upper and lower rates are often fixed to maximum and minimum rates by introducing a cap and a collar, respectively. The sponsor is protected from interest rate increases over the agreed cap, and the lender from decreases below the agreed collar.

In some cases the rate of interest will be determined as

Rate of interest = Relevant margin + Libor

where Libor is the arithmetic mean of the rates quoted by the reference banks.

An example of applicable margins are:

❑   Main loan facility
  ■ pre-completion 1.25%
  ■ post-completion 1.00%

❑ standby facility
  ■ pre-completion 1.5%
  ■ post-completion 1.25%
❑ working capital facility 1.5%

*Security*

On conventional loan finance, the lender seeks to limit risk by insisting that the borrower provides security for the loan, this is usually in the form of a charge on the project's assets or through a guarantee. In projects where the amount involved may be too large for participants to provide a guarantee, non-recourse loans are normally adopted. Under this form of loan, the lender accepts that surplus cashflow over the operating costs will cover interest payments arising from the debt, or in some cases the facility itself has a realisable capital value.

Another type of finance that may be considered for projects, especially in developing countries, is countertrade. Under this form of agreement, goods are exchanged for goods. An example of this form of agreement is the use of oil as payment for goods and services supplied. In such a project the oil, for example, would be sold on the open market or under a sales contract to generate revenues for repayments of capital and interest.

*Debentures*

A debenture is a document issued by a company in exchange for money lent to that company. The company agrees to pay the lender a stated rate of interest, and also to repay or redeem the principal at some future date. Debentures can be traded in the same way as shares. The interest paid to debenture holders is deductible when calculating taxable profits, unlike dividend payments.

Other forms of lending may include export credits, a floating interest rate loan, currency swaps and revenue bonds.

In most cases the success of a project depends entirely on cash flow. Mezzanine finance is an innovative form of finance to enhance the return to shareholders. Mezzanine finance may be in the form of fixed- or floating-rate loans with second charge assets. These are similar to conventional loans in that they provide for the payment of interest and principal through a flexible amortisation schedule.

Alternative finance packages may be based on:

❑ royalty agreements, i.e. an agreement providing the lender with an agreed percentage of future revenues;
❑ unsecured loan stock, i.e. fixed-interest loan stock which gives the

lender the right to a fixed return and to obtain repayment of the principal at the end of a stated period;

❑ convertible unsecured loan stock, i.e. fixed-interest loan stock with the right to convert to ordinary share capital at a future date;

❑ redeemable preference shares, i.e. shares giving the investor the right to a fixed return and to obtain repayment of the investment at the end of a stated period;

❑ convertible preference shares, i.e. shares giving the investor the right to a fixed return and to convert to ordinary share capital at a future date.

A typical project loan facility may include:

❑ a main bank facility
❑ a letter of credit facility
❑ standby facility

The first two facilities are available to finance any cost associated with the acquisition, operation and maintenance, and costs associated with an existing facility included under the concession contract. They exist to finance the costs of, and associated with, loan facilities which form part of the finance plan of the project, and to finance the construction of the project.

The third facility will be available to meet any required drawdowns in excess of the main bank facility due to higher interest rates, lower than expected revenues in the early phase of operation, and additional works performed under the construction contract.

In addition to the loan facilities, a typical finance package may also include the payment of fees such as:

❑ an underwriting fee, payable as a percentage of the bank loan facility;
❑ a management/front-end fee, payable as a percentage of the total facilities put in place;
❑ a commitment fee, calculated on the undrawn balance;
❑ an agency fee, which is an annual fee paid to the lead lending bank;
❑ a success fee, payable as a percentage of the total loan once all loans have been secured.

### Equity finance

Equity finance is usually an injection of risk capital into a company or venture. Providers of equity are compensated with dividends from profits if a company or venture is successful, but with no return should the venture be loss-making. Debt service in the majority of cases takes first call on profits whether or not profits have been generated; dividends are paid after debt claims have been met. In the event of a company or venture becoming insolvent, equity investors rank last in the order of repayment and may lose their investment.

The amount of equity provided is considered to be the balance of the loan required to finance the project. The total finance package is often described in terms of the debt/equity ratio. In projects considered to have a large degree of risk, then a larger proportion of equity is normally provided.

The advantages of an equity investment are that equity may be used as a balancing item to accommodate fixed repayments, and that equity investors are often committed to the success of a project since they are organisations involved in the realisation of the project. Providers of equity fall into two categories, those with an interest in the project (contractors, vendors, operators) and pure equity investors (shareholders).

Sources of equity include public share issues, financial institutions such as pension funds, companies and individuals, participants such as constructors, suppliers, operators, vendors and government, and international agencies such as the International Finance Corporation (IFC), the European Bank for Reconstruction and Development (EBRD) and the European Investment Bank (EIB).

In a number of overseas projects, finance is often provided in a mixture of currencies, often based at the time of the loan on the host country's currency. The host country's inflation may often result in additional loans being required to complete a project if no fixed exchange rate is agreed. Fluctuations between the values of the currencies provided in a project loan need to be addressed by the lender.

Similarly loans to countries which do not have transferable or exchangeable currencies, such as those of many eastern European and African countries, need to be based on either guarantees from a third party or countertrade. In the case of the latter, a product is provided in lieu of the loan repayments. This may be in the form of minerals or services. However, should the selling price of such minerals or services reduce during the repayment period, then the lender may not even recoup the principal loaned. Lenders will often overcome this commercial risk by entering into a sales contract for the product or service at prices suitable to meet the loan schedule.

## 6.8 Appraisal and validity of financing projects

The financial viability of a project over its life-time must be clearly demonstrable to potential equity investors and lending organisations. In assessing the attractiveness of a financial package, project sponsors should examine the following elements:

❏ interest rate, debt/equity ratio (percentage being financed);
❏ repayment period, currency of payment, associated charges (legal, management and syndication fees), securities (guarantees from lenders) and documentation (required for application, activation and drawdown of loans).

Three basic financial criteria need to be considered in projects, i.e. finance must be cost-effective, so far as possible the skilled use of finance at fixed rates to minimise risks should be adopted, and finance would be required over a long term.

The project must have clear and defined revenues that will be sufficient to service principal and interest payments on the project debt over the term of the loans, and to provide a return on equity which is commensurate with development and the long-term project risk taken by equity investors. The EIB will normally fund infrastructure projects for a period of 25 years, and industrial/process plants up to 14 years. Institutional investors such as insurance companies and pension funds often consider projects with fixed rates of return up to 20 years to match the cash-flow characteristics of their liabilities. Lenders often refer to a robust finance package as one which will allow the repayment of loans should interest rates increase during the operation period.

The commercial and financial considerations, rather than the technical elements, are normally the determinants of a successful project. The political risk, which in turn influences the financial risk, is less controllable than the technical risks, which are often allocated to the constructor or operator.

When selecting the sources and forms of capital required, the following aspects should be considered: the strength of the security package, the perception of the country's risks and limits, the sophistication of local capital markets.

It is possible that project finance is provided entirely from debt, e.g. in the Dartford River crossing project. In such a case, the lenders take the risk that revenues will be sufficient to pay off the debt by the end of the concession period. In this project, for example, the promoter provided pinpoint equity of only £1000, a form of equity under which shareholders do not receive dividends and the remaining finance is through loans.

In some cases government organisations may take an equity participation in the project, but many sponsors believe the inclusion of the government among the project company shareholders can lead to bureaucratic interference with project development and operation which privatisation is supposed to avoid.

In build–own–operate–transfer (BOOT or concession) project strategies, for example, the providers of finance are compensated solely from the project revenue with recourse to the revenue stream. The payment of revenues in BOOT projects may be arranged according to the time-period revenue streams or user revenue streams.

A time-period revenue stream is normally a predetermined payment by the promoter to the sponsor which is independent of the usage of the facility, but which guarantees the promoter a secured income. For example, a promoter may lease a tunnel from a sponsor based on an annual payment irrespective of the number of vehicles using the facility.

However, in most cases user revenue streams are related to the level of use of the facility; the number of vehicles using a tunnel multiplied by the toll rate would be considered as a user revenue.

One of the most important elements to be met in project finance is how to provide security to non-recourse or limited-recourse lenders. If a sponsor defaults under a particular project strategy utilising a non-recourse finance package, the lender may be left with a partly completed facility which has no market value. To protect lenders, therefore, various security devices are often included. These may include:

❑ revenues which are collected in one or more escrow accounts maintained by an escrow agent independent of the sponsor company;
❑ the benefits of various contracts entered into by the sponsor, such as construction contracts, performance bonds, supplier warranties and insurance proceeds, will normally be assigned to a trustee for the benefit of the lender;
❑ lenders may insist upon the right to take over the project in cases of financial or technical default prior to bankruptcy, and bring in new contractors, suppliers or operators to complete the project;
❑ lenders and export credit agencies may insist on measures of government support such as standby subordinated loan facilities, which are functionally almost equivalent to sovereign guarantees.

In summary, the successful elements required in funding projects should include limited and non-recourse credit, debt financing entirely in local currency, equity finance in currencies considered to be relatively strong, major innovations in project financing, project creditors who are confident, and governments which accept some project risks and provide limited resources.

## 6.9  Risks

The identification of the risks associated with any project is a necessary step before analysis and allocation, especially in the early stage of project appraisal. Lenders and investors will only be attracted to projects that provide suitable returns on the capital invested.

A project often has a number of risks, namely, identifiable risks which are within the control of one or more of the parties to the project, risks which may not be within any parties reasonable control, but may be insurable at a cost, and uninsurable risks.

By identifying risks at the appraisal stage of a project, a realistic estimate of the duration and final costs and revenues of a project may be determined.

### Financial risks

Financial risks are associated with the mechanics of raising and delivering finance and the availability of adequate working capital. Financial risks may include foreign exchange risks, off-take agreements, take or pay terms, the effect of escalation and debt service risk. This may arise during the operation phase when the machinery is running to specification, but does not generate sufficient revenue to cover operating costs and debt service.

Financial risks may be summarised as:

❑ *interest*, i.e. type of rate (fixed, floating or capped), changes in interest rate, existing rates;
❑ *payback*, i.e. loan period, fixed payments, cash-flow milestones, discount rates, rate of return, scheduling of payments;
❑ *loan*, i.e. type and source of loan, availability of loan, cost of servicing loan, default by lender, standby loan facility, debt/equity ratio, holding period, existing debt, covenants;
❑ *equity*, i.e. institutional support, take-up of shares, type of equity offered;
❑ *dividends*, i.e. time and amounts of dividend payments;
❑ *currencies*, i.e. currencies of loan, ratio of local/base currencies, mixed currencies.

### Revenue risks

The risks associated with revenue generation are often considered on the basis of meeting demands, and may include:

- *demand*, i.e. the accuracy of demand and growth data, the ability to meet increase in demand, demand over the concession period, demand associated with existing facilities;
- *toll*, i.e. market-led or contract-led revenues, shadow tolls, toll level, currencies of revenue, tariff variation formula, regulated tolls, take and/or pay payments.
- *developments*, i.e. changes in revenue streams from associated developments.

### Commercial risks

Commercial risks are considered when determining the commercial viability of a project. Commercial risks affect the market and revenue streams, which may include risks associated with access to new markets, the size of existing markets, pricing strategy and demand. Commercial risks arising from a deterioration of the competitive market position and overly optimistic appraisals of the value of pledged securities, such as oil and gas reserves.

Project revenues can be either market tied or contract tied, and market-tied revenues normally impose higher risks on the project sponsor. Demand risks are normally uncontrollable on a road project, and if demand is less than predicted then debt service may become impossible. Market risks such as the failure of customers to meet sales agreements, price fluctuations, inaccurate assessments of outlets, market uncertainty, obsolescence of the product or cancellation of long-term purchasing contracts may result in the commercial failure of a project. One way to minimise product market risk is direct financial participation by customers.

Market risks prior to completion of a project would normally include:

- raw materials not available when required during the construction phase;
- the market price of raw materials increases during the construction phase;
- the market price of the product falls, or fails to rise as anticipated, during the construction phase;
- market forces change, leaving no receptive market for the product once production commences.

In international projects, foreign exchange risks may be minimised by financing a project in the same mixture and proportion of currencies as those anticipated from the revenue streams. Cash flows determined by a number of currencies may be simplified by classifying currencies into groups. The two major base currencies used in this type of simplification are the US dollar and the German mark.

Commercial risks may be summarised under six main headings.

❑  Market: i.e. changes in demand for a facility or product, escalation of costs of raw materials and consumables, recession, economic downturn, quality of product, social acceptability of user-pay policy, marketing of product and consumer resistance to tolls.
❑  Reservoir: i.e. changes in input source.
❑  Currency: i.e. convertibility of revenue currencies, fluctuation in exchange rates, devaluation.

Lenders and investors should evaluate these risks to determine the commercial viability of a project. Should a project seem to be commercially viable from a financial point of view, then additional risk areas associated with construction, operation and maintenance, and political, legal and environmental factors should be identified and appraised.

❑  Sensitivity: i.e. location of project, existing environmental constraints, impending environmental changes.
❑  Impact: i.e. effect of pressure groups, external factors affecting operation, effect of environmental impact, changes in environmental consent.
❑  Ecological: i.e. changes in ecology during concession period.

A typical project finance analysis can be summarised under four main headings.

❑  Financial market analysis, which considers data regarding the availability, cost and conditions of financing a project.
❑  Cost analysis, which estimates the development, construction and operating costs and establishes a minimum cost of the project.
❑  Market analysis, which forecasts demand and establishes a maximum price, and also evaluates the commercial viability of the project.
❑  Financial analysis, which compares the cost, the market and the financial market analysis and establishes the relationship between costs and revenues.

Practical methods of reducing risks in international investment analysis may include:

❏ reducing the minimum payback period;
❏ raising the required rate of return of the project investment;
❏ adjusting cash flows for the cost of risk reduction;
❏ adjusting cash flows to reflect the specific impact of a particular risk.

## *Allocation of risks*

Typical responses by lenders to minimising risks in projects may be summarised under the headings listed below.

❏ Completion risk: cover by a fixed price, firm date, turnkey construction contract with stipulated liquidated damages.
❏ Performance and operating risk: cover by warranties from the constructor and equipment suppliers, and performance guarantees in the operation and maintenance contract.
❏ Cash-flow risk: cover by utilising escrow arrangements to cover forward debt service and guard against possible interruptions, and take out commercial insurance.
❏ Inflation and foreign exchange risk: cover by government guarantees regarding tariff adjustment formula, minimum revenue agreements and guarantees on convertibility at certain agreed exchange rates.
❏ Insurable risks: cover by a form of insurance such as a policy to cover cash flow shortfalls, mainly during the pre-completion phase of a project.
❏ Uninsurable risks: cover by insisting host government provide some form of coverage for uninsurable risks such as force majeure.
❏ Political risks: cover by political risk insurance from export credit agencies or multilateral investment agencies.
❏ Commercial risk: cover by insurance policies such as the Export Credit Guarantee Department (ECGD).

## References

Merna, A. and Owen, G. (1998) *Understanding the Private Finance Initiative. The New Dynamics of Project Finance*, Asia Law & Practice, Hong Kong, p. 78.
Nevitt, P.K. (1983) *Project Finance*, 4th edn, Bank of America, Financial Services Division.

## Further reading

Jackson, M. and Richardson, M. (1990) *Financing Construction Projects: Financial Control*, Engineering Management Guide, Series (ed. M. Barnes) Thomas Telford, London, pp. 61–2.

Merna, A. (1998) A Financial Risk in the Procurement of Capital and Infrastructure Projects, *International Journal of Project and Business Risk Management*, **2**(3).

Merna, A. and Dubey, A. (1998) *Financial Engineering in the Procurement of Projects*, Asia Law & Practice, Hong Kong.

Merna, A. and Njiru, C. (1998) *Financing and Managing Projects*, Asia Law & Practice, Hong Kong.

Merna, A. and Smith, N.J. (1996) *Projects Procured by Privately Financed Concession Contracts*, Vol. 1, 2nd edn, Asia Law & Practice, Hong, Kong.

Merna, A. and Smith, N.J. (1996) *Projects Procured by Privately Financed Concession Contracts*, Vol. 2, 1st edn, Asia Law & Practice, Hong Kong.

Merna, A. and Smith, N.J. (1996) *Guide to the Preparation and Evaluation of Build–Own–Operate–Transfer (BOOT) Project Tenders*, 2nd edn, Asia Law & Practice, Hong Kong.

NEDO (1990) *Private Participation in Infrastructure*, National Economic Development Council, London.

# Chapter 7

# Cost Estimating in Contracts and Projects

The purpose of this chapter is to indicate the theory, techniques and practical implications of cost estimating throughout the stages of the project cycle.

## 7.1 Cost estimating

The record of cost management in the engineering industry is not good. Many projects show massive cost and time overruns. These are frequently caused by underestimates rather than by failures of project management.

Estimates of cost and time are prepared and revised at many stages throughout the development of a project (Figure 7.1). They are all predictions, i.e. the best approximation that can be made, and it would be unrealistic to expect them to be accurate in the accounting sense. The degree of realism and confidence achieved will depend on the level of definition of the work and the extent of risk and uncertainty.

The objective of any estimate, no matter at what stage it is produced, is to predict the most likely cost of the project. It is essential to recognise that there is a range of possible costs within which the most likely cost lies. If the limits for the possible range of project costs over its timescale are plotted, an envelope similar to that in Figure 7.2 is shown.

The key points to note are given below.

- ❑ Note the range of possible distributions. Generally, these show a narrowing range and an increasing certainty as the project progresses, but certainty is only achieved following settlement of the final account(s) and auditing of all project-related costs.
- ❑ The range of possible costs is much greater during the early stages. This arises because there is little information available to the estimator in terms of scope, organisation and time, and cost data.

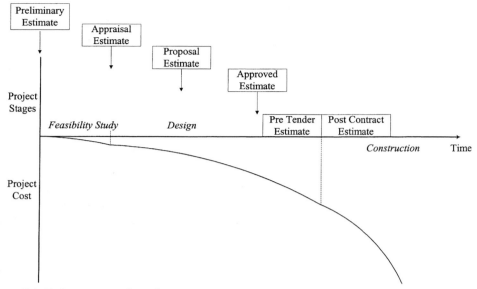

**Figure 7.1** Estimates at each project stage.

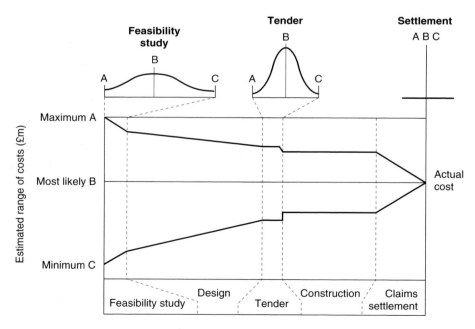

**Figure 7.2** Range of estimates over the project cycle.

❑ Many risks are latent in a project at its earliest stages. These often go unrecognised, because the project team is intent on looking ahead towards the project's completion rather than reviewing the whole of its evolution.

❑ Risk decreases over the life of a project. This may not be continuous; from time to time there may be increasing risks or new risks which arise during the project's development. Some of these are the result of a failure to recognise their importance or presence earlier.

❑ The upper and lower bounds of the envelope are not absolute, and in reality are not clearly defined.

❑ The estimated base cost plus contingencies which are likely to be spent must be close to the 'most likely' cost from the earliest stage if the common experience of underestimation is to be avoided. It follows, therefore, that the most likely value is closer to the maximum than the minimum to allow for cost growth.

Ideally, any estimate should be presented as a most probable value, with a tolerance, i.e. a range of less likely values to emphasise that it is an estimate. Particular areas of risk and uncertainty should be noted and, if necessary, a specific contingency allowance or tolerance should be included in the estimate.

The requirements of an estimate are:

❑ to predict the most probable cost of the works;
❑ to predict the most probable programme for the works;
❑ to identify and quantify potential problems and risks;
❑ to state all assumptions and note all exclusions; and
❑ to base a forecast of expenditure, i.e. the cash flow based on the project programme.

The most important estimates for a project are probably those prepared at sanction, and for a contract, those prepared at tender. It is at these points that the promoter and contractor become committed.

## 7.2 Cost and price

*Cost* refers to the cost directly attributable to an element of work, including direct overheads, e.g. supervision.

*Price* refers to the cost of an element of work, plus an allowance for general overheads, insurances, taxes, finance (and interest charges) and profit (sometimes known as 'the contribution').

First, the most probable cost of the works must be ascertained.

The traditional view in construction projects, as exemplified by the 'schedule of rates' or 'bill of quantities' (BoQ) approach, is that costs are proportional to quantities. This is true only to a very limited extent.

In reality, the only truly quantity-proportional costs are the direct costs of materials in the permanent works, but care is required. For example, the cost of concrete may vary depending upon whether it is site-batched or not.

The cost of operating a hoist or crane is a time related cost. It requires an operator, maintenance, fuel and lubricants when it is working, regardless of the volume of concrete which it handles. The cost of setting up the hoist or crane and later dismantling it are lump sum start and finish, or on/off, costs.

Time-related costs are often dominant, but there are others which must be considered:

❑ finance charges on money borrowed for the purchase of plant and equipment;
❑ depreciation in the value of a piece of plant as it ages.

These are highest for new pieces of equipment, and decrease steadily until the debt is repaid or, as in the second case, until the piece of equipment has only a residual or scrap value.

Time depends upon the output or productivity which can be obtained from the resources. Outputs and productivities are expressed in terms of a quantity of work in unit time, or time for a given quantity of work. Hence, the quantity of work to be done, say in the BoQ, is related to the level of resource which must be provided to carry out the work in the given time. If the time taken to carry out the work is extended beyond the optimum then the cost will increase unless the provision of the resource is reduced. It is important to realise that in reality, major plant and equipment numbers and labour force sizes cannot be varied on a day-to-day basis, so the actual cost incurred is dependent upon the provision of the resource, regardless of the quantity of work which it performs.

Therefore, the cost of an element of engineering work comprises quantity-proportional material costs, time-related labour, plant and equipment costs, and lump-sum costs. The latter may be for on/off costs for a particular item, or a payment to a specialist sub-contractor. These costs refer to the permanent works. Any temporary works must also be covered by the prices charged by the contractor (or sub-contractor). Finally, the direct overhead costs of management, establishment and consumables which can be charged against a project can be assessed.

These must then be spread over the other cost centres to arrive at the total cost to perform the work. The price of the work is derived from the cost. It is sometimes described as cost plus mark-up. This is a gross over-simplification.

There are general overheads incurred by an organisation which cannot be charged directly to a particular project, for example administrative staff, senior management, the cost of maintaining the head office buildings, and insurances. The price must also cover the payment of taxes, interest charges on monies borrowed, and allowances or contingencies for risks and uncertainties. Finally, the contractor must make a profit after tax.

The amount of profit which can be made varies. Some factors which affect it are:

❑ the type of work;
❑ the size of the project or contract;
❑ the extent of competition;
❑ the desire of the organisation to secure the work.

These may be grouped together very loosely as 'the state of the market'. This is possibly the factor influencing the price which is the most difficult to assess, and experience, good judgement, knowledge of the market and not a little luck are required to make a realistic assessment.

Other factors which influence the overall cost, and hence price, of engineering works must not be overlooked, these may include:

❑ the location;
❑ the degree of innovation;
❑ the type of contract;
❑ the method of measurement;
❑ the payment conditions;
❑ the risk surrounding the project.

## 7.3 Importance of the early estimates

All estimates should be prepared with care. The first estimate that is published for review and approval has a particularly crucial role to play because:

❑ it is the basis for releasing funds for further studies and estimates;
❑ it becomes the marker against which subsequent estimates are compared.

There is a further reason for the importance of early estimates; i.e. the need to know, for the purposes of economic appraisal, the capital cost of the project. The capital cost may not be the most sensitive variable in such an appraisal, but it can be decisive in the key decision to proceed with a project or to invest the funds elsewhere.

The decision of the promotor(s) to sanction the project should never be based upon the very first estimate. Nevertheless, before significant monies are spent on developing designs and more detailed estimates, there is a need to have a realistic indication of the project's likely cost. By definition, this must be done before a great deal of detailed information is available.

The earliest estimates are primarily quantifications of risks: there is little reliable data available to the estimator. To estimate effectively therefore requires that the estimator not only has access to comprehensive historical data, and is capable of choosing and applying the most appropriate technique, but that they also have, in conjunction with other members of the project team, the experience to make sound judgements regarding the levels of largely unquantifiable risk. For many projects, this will not be an onerous task, but for projects which are in any way unusual it is essential that thorough exercises in risk identification, risk assessment and the selection of the most appropriate responses are performed. In so far as cost estimates are concerned, the latter is the quantification of allowances for uncertain items, specific contingencies and general estimating tolerances. In addition, the project's programme must be carefully reviewed, the costs of delay, and the use of float or acceleration must be assessed, and the appropriate contingencies must be included in the estimate.

It is easy to see that this can be a time-, resource- and cost-intensive exercise. However, it is not necessary for all projects. The first step towards assessing the risk is to identify potentially high-risk projects. This can be achieved by considering a number of factors, regardless of the size, complexity, novelty or value of the particular project:

❏ the sensitivity of the promoter's business or economy to the outcome of the project in terms of the quality of its product, capital cost and timely completion;
❏ the need for new technology or the development of existing technology;
❏ any major physical or logistical restraints such as extreme ground conditions or access problems;
❏ novel methods of implementation;
❏ large and/or extremely complex project;

❑ any extreme time constraints;
❑ location; e.g. the parties involved (promoter, consultants and contractors) might be inexperienced;
❑ sensitivity to regulatory changes;
❑ developed or developing country.

Testing new projects against such criteria is the first step towards improving the realism of initial estimates.

If the result of the test outlined above is positive, that is, the project is inherently risky, further and detailed assessments must be performed of the potential sources of risk and their likely impact on the project.

There are two components to any estimate:

❑ the base estimate;
❑ contingency allowances which are required to cover the uncertainty in the base estimate.

Next, a list of the main sources of risk should be prepared. This may be as few as five, but should not exceed fifteen. For example:

❑ promoter
❑ host government(s)
❑ funding
❑ definition of project
❑ concept and design
❑ local conditions
❑ permanent installed plant (mechanical and electrical)
❑ implementation
❑ logistics
❑ estimating data (time and cost)
❑ inflation
❑ exchange rates
❑ force majeur

It should be noted that these are both quantitative and qualitative. Following this, each source of risk should be developed into several comprehensive lists of more specific, manageable and, where possible, quantifiable risk factors. It is useful at this stage to classify both the potential impact, major or minor, and the probability of occurrence, high or low, as a guide to the most important factors, because the estimator's attention should mainly be directed at these high-impact high-probability risks.

## 7.4  Estimating techniques

Having distinguished between cost and price, it is necessary to consider the estimating function in greater depth. The five basic estimating techniques available to meet the project needs outlined above, together with the data required for their application, are summarised in Table 7.1.

**Table 7.1**  Data for the basic estimating techniques (Smith, 1995).

| | Estimating technique | | | | |
|---|---|---|---|---|---|
| | Global | Factorial | Man-hours | Unit rate | Operational |
| Project data required | Size/capacity Location Completion date | List of main plant items installed Location Completion date | Quantities Location Key dates Simple method statement Completion date | Bill of quantities (at least main items) Location Completion date | Materials quantities Method statement Programme Key dates Completion date |
| | List of potential problems, risks, uncertainties and peculiarities of the project | | | | |
| Basic estimate data required | Achieved overall costs of similar projects (adequately defined) | Established factorial estimating system Recent quotes for main plant items | Hourly rates Productivities Overheads Materials costs | Historical unit rates for similar work items Preliminaries | Labour rates and productivities Plant costs and productivities Materials costs Overhead costs |
| | Inflation indices Market trends General inflation forecasts | Inflation indices (for historial prices) Market trends General inflation forecasts | Hourly rate forecasts Materials costs forecasts Plant data | Inflation indices Market trends General inflation forecasts | Labour rate forecasts Materials costs forecasts Plant capital and operating costs forecasts |

Three of the techniques, global, factorial and unit rates, rely on historical cost and price data of various kinds. Comments on this aspect of each technique are given under the respective headings, but the associated risks are so important that it is worth making the following general warning points about the use of historical data in estimating.

(1)  To obtain realistic estimates, the data must be from a sufficiently large sample of similar work in a similar location and constructed in similar circumstances.

(2)   Cost data need to be related to a specific base date. In the case of construction work carried out over a period of time, a 'mean' date has to be chosen, for example two-thirds of the way through the period. In the case of manufactured plant the easiest date to determine will probably be the delivery date, whereas a 'mean' date during manufacture may be more relevant.

(3)   Having selected the relevant base date, there remains the problem of updating the cost price data to the base date for the estimate. The only practical method is to use an inflation index, but there may not be a sufficiently specific index for the work in question. If there is not, recourse to general indices is usually made. In any event, there is a limited length of time over which such updating has any credibility, particularly in times of high inflation and/or rapidly changing technology.

(4)   Overlying the general effect of inflation is the influence of the 'market'. This will vary with the type of project being undertaken, and for international projects with the host country and also with the supplying countries. The state of the market at the price date base will require careful consideration before historical data can be credibly applied to later or future dates.

*Global*

This term describes the 'broadest brush' category of techniques which relies on libraries of achieved costs of similar projects related to the overall size or capacity of the asset provided. This technique may also be known as 'order of magnitude', 'rule of thumb' or 'ballpark' estimating. Examples are:

❑   cost per square metre of building floor area or per cubic metre of building volume;
❑   cost per megawatt capacity of power stations;
❑   cost per metre/km of roads/motorways;
❑   cost per tonne of output for process plants.

The technique relies entirely on historical data, and therefore must be used in conjunction with inflation and a judgement of the trends in price levels, i.e. market influence, to allow for the envisaged timing of the project.

The use of this type of 'rolled up' historical data is beset with dangers, especially inflation, as outlined generally above, but some more specific problems are listed below.

- ❑ Different definitions of what costs are included, for example:
  - ■ engineering fees and expenses by consultants/contractors/promoter, including design, construction supervision, procurement and commissioning;
  - ■ final accounts of all contracts, including settlements of claims and any '*ex gratia*' payments;
  - ■ land;
  - ■ plant and fittings supplied directly;
  - ■ transport costs;
  - ■ financing costs;
  - ■ taxes and duties.
- ❑ Different definitions of measurement of the unit of capacity, for example:
  - ■ a square metre of building floor area (is it measured inside or outside the external walls);
  - ■ a cubic metre of building volume (is the height measured from the top of the ground floor or the top of the foundations);
  - ■ a metre/kilometre of motorway (is this an overall average, including pro rata costs of interchanges, or should these be estimated separately);
  - ■ the units included in the associated infrastructure (for example, transmission links/roads for power stations).
- ❑ Not comparing like with like, for example:
  - ■ differing standards of quality, such as different runway pavement thicknesses for different levels of duty;
  - ■ process plants on greenfield sites and on established sites;
  - ■ different terrain and ground conditions, e.g. roads across flat plains compared with those across mountainous regions;
  - ■ different logistics depending on site location;
  - ■ scope of work differences, such as power stations with or without workshops and stores, housing compounds, etc;
  - ■ item prices taken out of total contract prices (especially turnkey) may be distorted by front-end loading to improve the contractor's cash flow, especially for hard-currency items.
- ❑ Inflation, e.g. different cost base dates:
  - ■ as noted above, it is essential to record the 'mean' base date for the achieved cost and use appropriate indices to adjust to the forecast date required.
- ❑ Market factors:
  - ■ competition for resources during periods of high activity;
  - ■ developing technology may influence unit costs.

Many of these items are obvious but for projects that do not have a continuous gestation period and may involve several different parties from time to time, it is essential that they are checked thoroughly.

A scrutiny of all these dangers, especially the effects of inflation, must be made before any reliance can be placed on a collection of data of this type. It follows that the most reliable data banks are those maintained for a specific organisation where there is confidence in the management of the data. The wider the source of the data, the greater is the risk of differences in definition.

However, as long as care is taken in the choice of data, the global technique is probably as reliable as an over-hasty estimate assembled from more detailed unit rates drawn from separate, unrelated sources and applied to 'guesstimates' of quantities.

## Factorial

These techniques are widely used for process plants and power stations, and also where the core of the project consists of major items of plant which can be specified relatively easily and current prices can be obtained for them from potential suppliers. The techniques provide factors for a comprehensive list of peripheral costs such as pipework, electrical work, instruments, structure or foundations. The estimate for each peripheral will be the product of its factor and the estimate for the main plant items.

The technique does not require a detailed programme, but nevertheless one should be prepared to identify problems of implementation, and lead times for equipment deliveries or planning approval which will go undetected if the technique is applied in a purely arithmetical way.

The success of the technique depends to a large extent on four factors.

(1)   The reliability of the budget prices for the main plant items. The estimator is still required to make a judgement on the value to include in his estimate depending on the state of the market and the firmness of the specification.

(2)   The reliability of the factors, which should preferably be the result of long experience of similar projects by the estimator's organisation. During periods of significant design development, certain factors can change rapidly (e.g. instrumentation and controls systems).

(3)   The experience of the estimator in the use of the technique and his ability to make relevant judgements.

(4)   The adoption of the technique as a whole, so that deficiencies in some areas will compensate for excesses in others.

The technique has the considerable advantage of being predominantly based on current costs, thereby taking account of market conditions and needing little, if any, reliance on inflation indices.

Factorial techniques are not normally reliable for site works, including most civil, building, mechanical and electrical installation work, except in a series of projects where the site circumstances are closely similar. In most overseas locations, the site works would need to be estimated separately using a more fundamental technique such as the operational one.

### Man-hours

Estimating the man-hours required is probably the original estimating technique. It is most suitable for labour-intensive implementation and construction, and operations such as the fabrication and erection of piping, mechanical equipment, electrical installations and instrumentation; work where there exist reliable records of the productivity of different trades per man-hour (for example process plant construction and fabrication of offshore installations). The total man-hours estimated for a given operation are then costed at the current labour rates and added to the costs of materials and equipment. The advantages of working in current costs are obtained.

The technique is similar to the operational technique. In practice, it is often used without a detailed programme on the assumption that the methods of construction will not vary from project to project. Experience has shown, however, that where they do vary (for example, the capacity of heavy lifting equipment available in fabrication yards), labour productivities, and consequently the total cost, can be significantly affected. It is recommended that a detailed programme is prepared when using this technique to identify constraints peculiar to the project. The prediction of cash flow requires such a programme.

### Unit rates

This technique is based on the traditional BoQ approach to pricing construction work. In its most detailed form, a bill will be available containing the quantities of work to be constructed, measured in accordance with an appropriate method of measurement. The estimator selects historical rates or prices for each item in the bill using either information from recent similar contracts, or published information (for example, price books for building or civil engineering), or 'built-up' rates from his analysis of the operations, plant and materials required for the item measured. As the technique relies on historical data, it is subject to the general dangers outlined above.

When a detailed bill is not available, quantities will be required for the main items of work, and these will be priced using 'rolled up' rates which take account of associated minor items. Taken to an extreme, the cruder unit rate estimates come into the area of global estimates, as described above (for example, unit rate per metre of motorway).

The technique is most appropriate to building and repetitive work, where the allocation of costs to specific operations is reasonably well defined and operational risks are more manageable. Sophisticated methods of measurement for building have been developed in the UK, and internationally many contractors are able to tender by pricing bills of quantities using rates based directly on continuing experience. Nevertheless, it is essential that the rates are selected from an adequate sample of similar work with reasonably consistent levels of productivity and limited distortions arising from construction risks and uncertainties, e.g. access problems.

The technique is less appropriate for civil engineering, where the method of construction is more variable and where the uncertainties of ground conditions are more significant. It is also likely to be less successful for both civil engineering and building projects in locations where few similar projects have been completed in the past. Its success in this area depends much more on the experience of the estimator and his access to a well-understood data bank of relevant 'rolled up' rates.

The unit rates quoted by contractors in their tenders are not necessarily directly related to the items of work they are pricing. It is common practice for a tenderer to distribute the monies included in his tender across the items in the bill to meet objectives such as cash flow and anticipated changes in volume of work, as well as taking some account of the bill item descriptions. It is unlikely that a similar 'weighting' is carried out by all tenderers in an enquiry, and therefore it is not easily detected. It follows that tendered bill unit rates are not necessarily reliable guides to prices for the work described.

From the promoter's point of view, the technique does not demand an examination of the programme and method of construction, and the estimate is compiled by the direct application of historical 'prices'. Therefore it does not encourage an analysis of the real costs of work of the kind that would need to be undertaken by a tendering contractor for any but the simplest of jobs. Neither does it encourage consideration of the particular peculiarities, requirements, constraints and risks affecting the project.

There is a real danger that the precision and detail of the individual rates can generate a misplaced level of confidence in the figures. It must not be assumed that the previous work was of the same nature, carried out in identical conditions and of the same duration. The duration of the work will have a significant effect on the cost. Many construction costs are time related, as are the fees of supervisory staff, and all are affected by inflation.

It is therefore recommended that a programme embracing mobilisation and construction is prepared. This should be used to produce a check estimate in a simplified operational form where there is any doubt about the realism of the unit rates available.

Despite its shortcomings, unit-rate estimating is probably the most frequently used technique. It can result in reliable estimates when practised by experienced estimators with good intuitive judgement, access to a reliable, well-managed data bank of estimating data, and the ability to assess the realistic programme and circumstances of the work.

### Operational cost (resource cost)

This is the fundamental estimating technique, since the total cost of the work is compiled from a consideration of the constituent operations or activities revealed by the method statement and programme, and from the accumulated demand for resources. Labour, plant and materials are costed at current rates. The advantage of working in current costs is obtained.

The most difficult data to obtain are the productivities of labour, plant and equipment in the geographical location of the project, and especially in the circumstances of the specific activity under consideration. Claimed outputs of plans are obtainable from suppliers, but these need to be reviewed in the light of actual experience. Labour productivities will vary from site to site depending on management, organisation, industrial relations, site conditions and other factors, and also from country to country. Collections of productivity information tend to be personal to the collector, and indeed this type of knowledge is a significant part of the 'know how' of a contractor and will naturally be jealously guarded.

The operational technique is particularly valuable where there are significant uncertainties and risks. Because the technique exposes the basic sources of costs, the sensitivities of the estimate to alternative assumptions or methods can easily be investigated, and the reasons for variations in cost appreciated. It also provides a detailed current cost and time basis for the application of inflation forecasts, and hence the compilation of a project cash flow.

In particular, the operational technique for estimating holds the best chance of identifying risks of delay as it involves the preparation of a method of implementation, and a sequential programme including an appreciation of productivities. Sensitivity analyses can be carried out to determine the most vulnerable operations, and appropriate allowances included. Action to reduce the effects of risks should be taken where possible.

This is the most reliable estimating technique for engineering work. Its execution is relatively time-consuming and resource-intensive compared with other techniques, but estimating organisations geared up to this technique accumulate data in an operational form which enables them to prepare even preliminary estimates with some appreciation of the more obvious risks, uncertainties and special circumstances of the project.

## 7.5  Suitability of estimating techniques to project stages

The objective should be to evolve a cost history of the project from inception to completion with an estimated total cash cost at each stage nearer to the eventual outturn cost. This can be achieved if the rising level of definition is balanced by reducing tolerances and contingency allowances, which represent uncertainty. Ideally, each estimate should be directly comparable with its predecessor in a form suitable for cost monitoring during implementation, and for a usable record in a cost data bank. In practise, this may be difficult to achieve. There is some correlation between the five estimating techniques which have been described and the estimating stages which have been defined. This is related to the level of detail available for estimating.

### *Preliminary*

This is an initial estimate at the earliest possible stages. There is likely to be no design data available, and only a crude indication of the project size or capacity, and the estimate is likely to be of use in provisional planning of capital expenditure programmes.

At this stage the global estimating technique can be used. This is a crude system which relies upon the existence of data for similar projects assessed purely on a single characteristic such as size, capacity or output. Widely used on process plants is the factorial method, where the key components can easily be identified and priced, and all other works are calculated as factors of these components.

*Feasibility*

Sometimes known as appraisal estimates, these are directly comparable estimates of the alternative schemes under consideration. They should include all costs which will be charged against the project to provide the best estimate of anticipated total cost, and if this is to be used to update the initial figure in the forward budget, then it must be escalated to a cash estimate.

A price can be described by the following equation:

$$Price = Cost\ estimate + Risk + Overheads + Profit + Mark\text{-}up$$

The size of the profit margin and the commercial decision making behind the selection of the percentage mark-up or mark-down are not discussed in detail here.

The basic cost estimate is the largest of all these elements, often accounting for more than 90% of the total price. Usually, the basic cost estimate is derived from the unit rate or operational assessment of the labour, plant, materials and sub-contract work required. Quotations are required for materials and sub-contractors. Typically, materials account for between 30% and 60% of a project's value; sub-contractors can typically account for between 20% and 40%.

The cost of the company's own labour is usually calculated either per hour, per shift or per week. The cost to the company of employing their own labour is greater than that paid directly to the employee. The elements in the calculation could include such factors as plus rates for additional duties, tool money, travel monies, National Insurance, training levies, employers and public liability insurance, and allowances for supervision.

*Erection or implementation*

Plant may be obtained for a contract either internally or externally. Quotations for the plant are therefore obtained from external hirers or from the company's own plant department. It is rare for UK contractors in the domestic market to calculate the all-in plant rate from first principles. This calculation is usually undertaken by the plant hire company.

Overheads (or on-costs) for the project could include allowances for site management and supervision, clerical staff and general employees, accommodation, general items and sundry requirements.

*Design*

This is an estimate for the scheme selected, which usually evolves from a conceptual design until the immediate pre-tender definitive design is completed.

A man-hours method is most suitable for labour-intensive operations such as design, maintenance or mechanical erection. The work is estimated in total man-hours and costed in conjunction with plant and material costs.

The decisions made in the design stage about the size, quality and complexity of a project have the greatest influence on the final capital costs of a project. As a design develops, more and more capital cost is committed on behalf of the promoter until at the tender stage, with the design complete or virtually complete, the promoter is committed to a high percentage of the tender value. Unless a re-design is undertaken, with the consequent loss of fees and time, the ability to save money whilst maintaining the original design concept is very limited. The design budget estimate should confirm the appraisal estimate and set the cost limit for the capital cost of the project.

*Implementation or construction*

This is an opportunity for a further refinement to reflect the prices in the contract awarded. This would require some redistribution of the money, e.g. in the BoQ in a unit-rate contract, and would contribute to more efficient management of the contract. The unit-rate method is a technique based on the traditional bill of quantities approach, where the quantities of work are defined and measured in accordance with a standard method of measurement.

## 7.6 Estimating for process plants

The most significant difference between estimating for a process plant project and estimating for other types of engineering project is that the base cost of the process estimate is derived from material and equipment suppliers, plant vendors, specialist contractors and sub-contractors. These components commonly account for about 80% of the total cost of the project. Consequently, the project estimator must ensure that firm quotations, together with guaranteed delivery dates and installation schedules, are confirmed with all suppliers and sub-contractors.

The main process plant contractor, the engineering contractor, usually carries out the design, procurement and management functions, which account for most of the remaining cost, i.e. about 20%. The first task is for the engineering contractor to estimate his own base costs, as accurately as possible, for the work to be undertaken at the detailed design, procurement, project management and site supervision stages of the project. Typically, the erection of process plants is labour-intensive, and hence this assessment is made using the man-hours estimating technique.

The engineering contractor's control philosophy must be aimed at minimising changes at the detailed engineering phase. The initial global estimate can be made by comparative means with similar types of plant, based on throughput. The estimate will have taken into consideration the principal quantities of the work, the items to be subcontracted, the materials and plant for which quotes are required, critical dates for actions by sub-contractors and suppliers, and whether any design alternatives should be pursued.

At this stage, the effects of layout and location are considered in the preparation of the base, or nett cost, estimate. Costs will be established for project and engineering services, which are usually done in-house, and any remaining quotes are substantiated and confirmed. It is important to have a good definition of the scheme, including the general arrangement, piping and instrument diagrams, equipment lists and materials specifications.

In the production of a nett cost, allowances and contingencies need to be considered. Allowances can be defined as costs added to individual estimated costs to compensate for a known shortfall in data, or to provide for anticipated developments. It is often the case that when firm quotes are replaced with contracts and orders after the successful tender bid, more details are available than was the case earlier. Contingency is an adjustment to the nett cost, and has to be considered no matter how much detailed work has been completed during the preparation of the estimate as there will undoubtedly be uncertainties which will affect the final cost. Contingency adjustments allow for the unknown elements, and also for any factors outside the control of the engineer contractor which are perceived to be likely to affect the final cost.

The cost of preparing a fixed-price estimate can be as high as 1–3% of the total cost, compared with 0.5–1% for building and civil engineering work, and it can take up to 6 months to prepare. Typically, a 3-month tender period for most major process plant projects is allowed by most promoters, which is really too short to enable all the requirements of the estimate to be completed.

## 7.7  Information technology in estimating

The estimating techniques described above were mainly developed as manual methods. The development of computer hardware, dedicated computer software packages and increased computer literacy amongst the professionals engaged in project management and estimating has resulted in the application of information technology (IT) tools to facilitate and assist in the estimating process.

As in manual estimating, the role of computer estimating will depend upon the user's requirements. Those of a promoter at the design stage will be very different from those of a contractor at tender. The data available for the production of an estimate will also be different. This section does not contain details of specific hardware or software products, which are well marketed and are almost continually being updated and improved, but focuses on some of the main functions that IT can perform in this process.

IT is concerned with the storage, transfer and retrieval of data, manipulation and calculation using data, and the presentation of output. The degree of complexity will vary, as will the operation of the user interface, ranging from data libraries to fully automated expert systems. Most computer software estimating packages incorporate a data library for storing data, a range of methods for manipulating direct costs to produce a price, the ability to update or alter any of the input data, and appropriate reporting formats.

The software packages attempt to be user-friendly by permitting data input or price build-up by the estimator in a way which closely resembles the normal way of working. The estimator has to spend less time performing routine calculations and arithmetical checks, whilst the basic method remains unchanged, but has more time for the use of judgement and experience where appropriate. It is important to note that a computer-based estimate is only as accurate as the input data, and that the use of a computer in itself will not necessarily increase the accuracy of the estimates. Indeed, if the data library is not kept up to date, or if it is applied without careful thought, then the estimate produced might suffer from a decrease in accuracy.

The application of IT cannot replace the role of the estimator in the project; nor is it intended to do so. Software packages are readily available to assist with most types of estimating technique, but there are also other advantages for the project manager. One of the recognised sources of errors and misunderstanding in project management has been identified as the linking interface between the feasibility study, the project estimate and detailed planning. Often, these stages were tackled by different people using different estimating methods with associated differing assumptions, and where the previous estimate was usually abandoned once that phase of the project had been completed. Current estimating software packages can be related to a single database and also be linked with other computer programs. This allows all the work done at the feasibility stage to be refined, modified and developed as the project progresses, hence retaining all the information within the system. Typical interfaces relate to measurement and valuation, to BoQs, to the valuation of variations, and to delays and disruptions. An estimating package can be linked with a planning package, and data can automatically be transferred, facilitating an understanding of the relationship between time and money for the project.

Expert systems are attracting considerable interest as potential aids to decision making. In an advisory role, they could provide the necessary assistance towards producing the cheapest cost estimate by allowing automated decision making to select resources to match different work loads. The 'expert' approach purports to provide tools for the effective representation and use of knowledge developed from experience, allowing the optimisation of resource selection. A computer simulation then begins with a preliminary questions module in which the user responds to a set of questions posed by the system. Estimating systems are evolving within an integrated environment as a decision-making or decision support.

## 7.8  Realism of estimates

The use of the word realism in this context, rather than accuracy, is important. As noted above, estimates are not accurate in the accounting sense, and the make-up of the total must be expected to change. The realism of estimates will greatly depend on the nature and location of the work, the level of definition of the project or contract, and particularly on the extent of the residual risk and uncertainty at the time, as discussed above. Studies have determined that a standard deviation of 7% was common for contractors' bid estimates in the UK process plant industry, but the performance of particular companies varied from 4% to 15%. The ranges of accuracy for high-risk projects, and in particular development projects, may be much greater.

The variation on individual items within the estimate is much greater, and consequently any system of cost control based on a comparison of actual and estimated cost must be of dubious value. It is preferable to use performance measurement systems based on an integration of the cost and the programme. These systems are based on the concept of earned value.

Simulation studies using these systems suggested that an improvement in estimating accuracy will produce a corresponding improvement in company performance. Many estimating problems can be addressed by adopting:

☐ a structured approach;
☐ choice of the appropriate technique;
☐ use of the most reliable data;
☐ consideration of the risks.

The improvement in performance that should be obtained is illustrated in Figure 7.3.

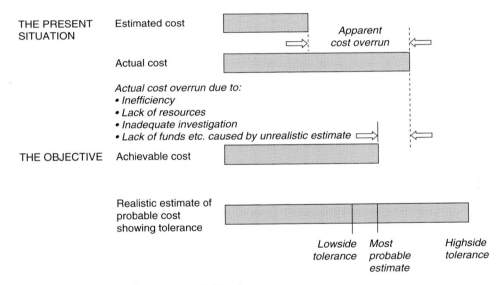

**Figure 7.3** Estimating performance and objectives.

## Further reading

Association of Cost Engineers (1991) *Estimating Checklist for Capital Projects*, 2nd edn, Association of Cost Engineers, London.

Bower, D., Thompson, P.A., McGowan, P. and Horner, R.M.W. (1993) Integrating project time with cost and price data. In: *Developments in Civil and Construction Engineering Computing*, 5th International Conference on Civil and Structural Engineering Computing, Civ-Comp Press, Edinburgh, pp. 41–9.

McCaffer, R. and Baldwin, A.N. (1991) *Estimating and Tendering for Civil Engineering Works*, 2nd edn, Blackwell Science, Oxford.

Smith, N.J. (ed.) (1995) *Project Cost Estimating*, Engineering Management Guide Series, Thomas Telford, London.

Sweeting, J. (1997) *Project Cost Estimating: Principles and Practices*, Institution of Chemical Engineers, Rugby.

Thompson, P.A. and Perry, J.G. (eds) (1992) *Engineering Construction Risks: A Guide to Project Risk Analysis and Risk Management*, Thomas Telford, London.

# Chapter 8
# Project Stakeholders

This chapter focuses on the interaction between the project and its external environment. Projects influence the environment, and will attract interest from people and organisations outside the project. These parties have the ability to either hinder or promote projects.

## 8.1 Stakeholders

All projects have to interact with the broader environment in which they exist. Any project that fails to take due cognisance of the impact it is having on the external environment is likely to face difficulties in its approval, planning or execution. Stakeholder analysis offers the opportunity to identify those parties that are influenced by the project or can influence the project. Edward Freeman (1984) suggests that a stakeholder in an organisation is any group or individual who can affect, or is affected by, the achievement of the organisation's objectives. This definition applied to a project implies that project stakeholders are those groups or individuals that can affect, or are affected by, the project. The project is open to a host of influences from stakeholders. Figure 8.1 provides a simple illustration of a stakeholder map.

## 8.2 Primary project stakeholders

The primary stakeholders on a project are those parties that have an immediate influence, or are influenced by, the project. They include, among others, the project's champions and sponsors, equity and debt holders, suppliers and contractors, and staff on the project. All these parties have an intrinsic interest in how the project performs, because they are critical to the very existence of the project. The success of failure of the project is dependent on how these primary stakeholders interact

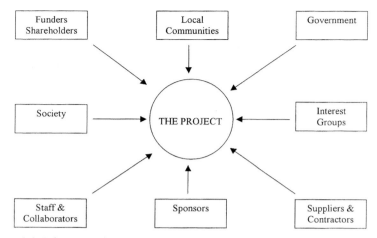

**Figure 8.1** Primary and secondary stakeholder map.

because they are integral to the project's processes, and value creation and the ability to achieve the goals of the project.

Projects are rarely stand-alone activities for primary stakeholders; they are part of a larger strategic business activity. The impact of a project may be such that its failure may affect the greater business activity. For instance, the failure of a new technology project or new product development project can reduce an organisation's competitiveness or market standing. It is important that the project management team understand what the project objectives mean for the on-going business activities of the project's champions, sponsors or financiers. This awareness is sometimes difficult to achieve, as some stakeholders are not keen to divulge information that is part of their competition strategy. In circumstances where the broader business objectives are communicated, there could still be uncertainty when the risks attached to the project are identified. However, it is important to note that not every primary stakeholder will be equally affected should a project be unsuccessful.

## 8.3 Secondary project stakeholders

The secondary group of stakeholders consists of those groups or individuals that are not directly related to the core of the project. They include groups such as the government, local authorities, unions, local communities, political parties, consumer groups, etc. There is a great deal of diversity and potential influence among these secondary stakeholders. Although they are not closely tied to the project, they can exercise a great deal of influence. For example, all levels of government can exert considerable influence on projects through policy, legislation and regulation. Tougher environmental controls with regard to waste disposal and management will have an increasing impact on engineering projects. Legislative and regulatory actions have an impact on all projects.

The impact of secondary stakeholders goes beyond just the realms of government. Projects have a direct impact on the environments in which they occur through their processes and end products. Visible engineering projects, or their by-products, are of interest to the public at large. This interest can either help promote or disrupt projects. Secondary stakeholder influence can appear very early in the life-cycle of the project during the feasibility and approval stages. All projects require approval at either local or national level. The process of approval often requires consultation in a public arena. These activities take time, and can have a knock-on effect on the project's success. The public enquiry on the proposed Terminal 5 facility at London Heathrow airport took 5 years. The cost of this enquiry is expected to be approximately £100 million. The scheme involves 11 local authorities and a variety of local, national and international interest groups who are all opposing it. It is expected that as a result of this process, there will be considerable difficulty in ensuring that the original cost targets are achieved. The inflationary impact of delays will require a rethink of the budgets. The original business assumptions have also changed with regard to airline travel, and this requires a redefinition of some aspects of the business case for the project. This high-profile project has met a great deal of local resistance, ensuring that a comprehensive enquiry was undertaken. This case demonstrates how projects can be frustrated if there is opposition on any major scale by secondary stakeholders.

The impact of secondary stakeholders can have varying degrees of influence. In some cases this may be supportive, but in the majority of cases it is disruptive. A key factor in this behaviour is that secondary stakeholders often have to live with the outcomes of the project over its whole life-cycle, rather than the limited period experienced by project managers or primary stakeholders.

## 8.4 Understanding the interests and influences

The use of the stakeholder concept is related to an identification of the interest, influence and impact that lies with each stakeholder. The use of the concept is directly related to the stake in the project, but this lacks tangibility and a degree of categorisation will help.

The easiest dimension is the economic influence on the project. This is where the stakeholders may have an influence on the profitability, cash flow or risk profile of the project. In turn, the project may have an influence on stakeholders through its activities, whereby it can influence market share, growth, prices, etc. Fortunes are often won or lost on the performance of a project. Suppliers, subcontractors, collaborators and joint venture partners would fall into both categories. The impact of external partners should not be underestimated as we adopt more cooperative approaches to project management.

Another dimension for particular stakeholders may be their technological influence on the project. Particular stakeholders may have the ability to prevent or assist the project to have access to technology, equipment or skills. This influence can potentially be strong if the technology is highly specialised and the stakeholder is in a monopoly situation. The use of specialist services are very prevalent in engineering, and as more activity is out-sourced the greater is the potential influence of highly skilled, technologically advanced subcontractors and suppliers. Changing technology can also add to the power of specialists.

Some stakeholders may have a social influence on the organisation, by altering its position in society, changing public opinion, or facilitating or hindering the company with their social influence. There is the potential for social stakeholders to act as a rallying point for other stakeholders. Consumer groups and environmental activists fall into this category. A good illustration relates to the decommissioning of the Brent Spar oil platform. The decommissioning project was technically feasible and cost-effective, but not socially acceptable. Greenpeace, as well as taking direct action to prevent the decommissioning, also galvanised opinion against the parent company. It acted as a rallying point for environmental interests and changed public opinion. Project managers are unlikely to be able to influence these situations unless they identify the potential impact of stakeholders early in the project's life. The influence of society is highly dependent on the values, morals and ethics present in any particular society. The value system of the sponsoring organisation, its projects and its stakeholders will define this relationship.

The social impact is very often the forerunner of the political effects of stakeholders. Social issues normally end up on the political agenda. Political influence followed by legislative capability can have a serious impact on projects. All forms of government have this type of influence, and may choose to exercise the influence to a greater or lesser degree. In certain cases the firm may have the ability to influence the political process. Politically and economically powerful organisations can use their influence to change political opinion in order to facilitate their projects. In a similar manner they can use this influence to disrupt or block projects. These stakeholders with political influence can redefine the potential impact of change on the firm. The political stake has to be considered carefully, particularly for projects of a sensitive nature and when operating internationally.

The management form of stake is often generated internally from the project organisation. This is the influence of the management philosophy and culture of the organisation, and it determines the systems, procedures, processes and values adopted by the company. In many ways this has a major impact on the relationships that exist between the project and its internal stakeholders. These relationships create the internal culture of the project. If a project operates in an environment of trust and cooperation, then it is likely that this will be continued externally. The culture of the project is a function of both its internal and external relationships. Collaborative working relationships benefit from having the appropriate culture within the project. If, for example, a project wants to operate in a non-adversarial manner, the staff being recruited should show that they can work in such a way and will fit into the culture of the project. The value system a company adopts internally continues externally, and therefore the internal values are important.

A categorisation of interests allows us to understand the nature of the relationship between the stakeholder and the organisation. It also provides a means of looking at cause and effect in the stake, and provides a mechanism by which to address any issues that arise. This cause and effect in the relationship is important for any form of proactive management of stakeholders.

## 8.5 Stakeholder management

The process of identifying potential stakeholders in a project should be a major part of the strategic management process of all projects. This is a problem for the project manager, as the number and nature of stakeholders on a project will vary with the life of the project. As the project evolves, then the stakeholders and stakes change. It would therefore make sense to carry out this identification throughout the life of the project. Taking a project as the equivalent of an organisation, it follows that stakeholder management should form a major part of the strategic management process at the start of a project. Stakeholder management is an important part of strategic project management.

Project stakeholder management is the process of dealing with the people who have an interest in the project, with the aim of aligning their objectives with those of the project. A systematic approach to the project stakeholder management process is shown in Figure 8.2.

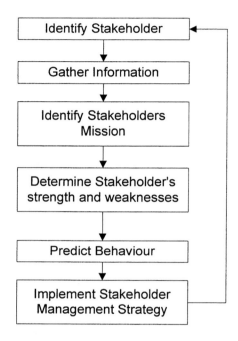

**Figure 8.2** Project stakeholder management process (adapted from Cleland and King, 1988).

The first step in the process is to identify who are the potential stakeholders on the project. Using simple stakeholder mapping and identifying primary and secondary stakeholders can achieve this objective. Creative thinking is needed to identify stakeholders who have no obvious interest in the project. A failure to identify the widest possible scope of stakeholders may generate problems later in the project.

The next step is to gather information about the stakeholders. This process is easy for public organisations, as they have information in the public domain that is easily accessible. Ad hoc or new organisations are difficult to research, particularly in a local context. Lack of information can lead to an underestimation of the potential impact of stakeholders. A small well-organised interest group may pose more problems if they have access to the media and political influence.

The following step is to identify what the stakeholder's mission for the project is likely to be. For example, if Greenpeace is identified as a potential stakeholder, gathering information about how they work and what aspect of the project is likely to attract their attention provides the basis for determining what their mission could be. At the heart of all these missions is the project itself and how it impacts on stakeholders. It is useful to remember that where the project team sees opportunities from the project, other interests see threats.

The next stage involves the determination of the stakeholder's strengths and weaknesses. This stage helps the management to understand how much the stakeholder can actually affect the project, what avenues they have available to make their needs known, and how far they will go to see them through. In such an analysis, it is best not to underestimate what stakeholders can achieve. The analysis of primary stakeholders may be easier as they want the project to succeed, but projects often run into problems by underestimating the capability of many secondary stakeholders. The best examples of this are environmental interest groups that have sophisticated organisations at local, national and international level. They have the ability to mobilise public opinion, have effective media management and good research departments, and can influence the political agenda, but they are often neglected in stakeholder analysis.

The next step is to predict what the behaviour of the stakeholder will be and how it will affect the project's goals and objectives. It is best to develop scenarios for a variety of situations that may arise from stakeholder actions. These scenarios can be used to establish risk profiles as well as contingencies for particular outcomes. Not all outcomes will be supportive or beneficial for the project, and it is sensible to consider all possible situations.

Once scenarios have been developed, the project manager can devise and implement stakeholder management strategies. In certain cases this would entail counteracting negative stakeholder behaviour. This may result in actions that force a redefinition or alteration of the project's objects, additional resource allocation, or simply in being more effective in communicating the project's objectives and its ultimate benefits. Even if stakeholder behaviour is supportive, it is still useful to remember that this interest in the project is important and provide the necessary support. Both disruptive and supportive behaviour should be considered.

The process of stakeholder management will follow the project's lifecycle, as the stakeholders in a project will vary from one stage in the life of the project to another, and strategies may sometimes have to be revised. As illustrated above, this systematic approach helps the project manager to manage proactively rather than reactively, as it raises awareness of the potential dangers of stakeholder action so that no-one is taken by surprise.

An example of how project managers identified the potential of external stakeholders and acted accordingly for the benefit of the project is the Prince Edward Island Bridge project. The project managers on the Prince Edward Island Bridge project in Canada did foresee delays and consequent cost overruns due to the opposition of the fishermen and ferry companies. To make the project more attractive to investors, the promoters obtained a security package to guard against delay and cost overruns. One of the things that the package had was a US$141 million performance bond. This package helped the project to survive delays resulting from court hearings and about 70 environment impact assessment studies that had to be carried out. In this case primary stakeholders helped provide solutions that counteracted opposition from secondary stakeholders.

Stakeholder management should lead to improved project objective achievement, while stakeholder neglect hinders it. The process of stakeholder management should also help to improve understanding of stakeholders and their motives.

## 8.6 Stakeholders and communication

The project manager is required to be an effective communicator and represent the interests of the project. Stakeholder identification shows that there is often a varied audience who is interested in the project. The project manager is now communicating to a wider audience. Communications to primary stakeholders will relate to progress, profitability, efficiency, timing and performance, as the primary stakeholder's interest is tied into the performance of the project. Strategies for communication with primary stakeholders will focus on their particular interests in the project, and may include personal briefings, newsletters, reports, meetings, workshops, and dedicated websites and intranets. The majority of primary stakeholders will look to the project manager to keep them informed about progress and performance.

Secondary stakeholders present a more complex communication challenge because of the variety of their interests in the project. Communication challenges may range from representing the project at a public inquiry to a newspaper interview about the project. Stakeholder management allows the project manager to identify the interests of secondary stakeholders and tailor communication to respond to these interests. It is also imperative that any communication includes the broader project objectives and benefits. Concentrating only on the key interest of stakeholders may lead to the overall objectives being obscured. The tools that a project manager will use for communication to secondary stakeholders will depend on the circumstances of the information exchange. These may vary from a presentation at a public meeting to a press release.

Communication skills have to include rebuttal techniques, as mis-information and negative communications will have to be managed. Stakeholders acting against the project may wage a negative information and media campaign. The internet has become the favourite medium for campaigns against projects. Projects can also be disrupted by direct action, protests, boycotts and strikes. The project management team has to be prepared for such situations. Stakeholder analysis should raise awareness of the likelihood of such situations and contingency plans should be prepared. There are situations where the project team may be faced with sophisticated and orchestrated campaigns against the project. In such circumstances, the project team should consider getting specialist help from either sponsors or external media specialists. As projects face greater scrutiny, project managers will have to acquire communication skills.

## 8.7  Summary

Projects do not operate in isolation; they are part of an open system. A project can change and influence its surroundings. Alternatively, the outcome of a project can be influenced by its surroundings. There is growing interest in projects because they have an impact on the broader economic, technological, ecological and social environment. This inter-est in a project may be either supportive or obstructive depending on how the project outcomes are likely to influence the environment. While supportive behaviour is encouraged, obstructive behaviour can present difficulties for the project. To gain better understanding of why there is interest in a project, it is possible to study how interests and influence may affect the project. A structured approach is required to consider these influences.

Stakeholder identification and management help to identify and plan for the various influences that may impinge on a project. Having a systematic approach to stakeholder management allows the project team to prepare plans to accommodate or counteract such influences and to communicate with the stakeholders. A structured approach also allows for a deeper understanding of the motivation of stakeholders and the actions they are likely to take. These could range from supporting the project to delaying or disrupting the project. As the power of stakeholders continues to rise, early planning is essential to manage this influence. Stakeholder identification and management recognises that projects operate in open systems, that external influences impinge on projects, and that projects also require an external orientation.

## References

Cleland, D.I. and King, W.R. (1988) *Project Management Handbook*, 2nd edn, Van Nostrand Reinhold, New York.

Freeman, R.E. (1984) *Strategic Management: A Stakeholder Approach*, Pitman, Marschfield, MA.

## Further reading

Kanter, R.M. (1995) *World Class: Thriving Locally in a Global Economy*, Simon and Schuster, New York.

Preece, C.N., Moodley, K. and Smith, A.M. (1998) *Corporate Communication in Construction: Public Relations Strategies for Successful Business and Projects*, Blackwell, Oxford.

# Chapter 9
# Planning

This chapter reviews some programming and planning techniques for the design, procurement, construction and commissioning of projects. The role of information technology is examined, together with suggestions on the identification and selection of appropriate software packages.

## 9.1 Planning

The successful realisation of a project will depend upon careful and continuous planning. The activities of designers, manufacturers, suppliers and contractors, and all their resources, must be organised and integrated to meet the objectives set by the promoter and/or the contractor.

The purposes of planning are to persuade people to perform tasks before they delay the operations of other groups of people, in such a sequence that the best use is made of available resources, and to provide a framework for decision making in the event of change. Assumptions are invariably made as a plan is developed; these should be clearly stated, so that everyone using the plan is aware of any limitations on its validity. Programmes are essentially two-dimensional graphs, and in many cases are used as the initial, and sometimes the only, planning technique.

Packages of work, usually referred to as 'activities' or 'tasks', are determined by consideration of the type of or location of the work, or by any restraints on the continuity of the activity. Activities consume 'resources', which are the productive aspects of the project and usually include the organisation and utilisation of people (labour), equipment (plant) and raw materials. Sequences of activities will be linked on a time-scale to ensure that priorities are identified and that efficient use is made of expensive and/or scarce resources within the physical constraints affecting the job.

A degree of change and uncertainty is inherent in engineering, and it should be expected that a plan will change. It must therefore be capable of being updated quickly and regularly if it is to remain a guide to the most efficient way of completing the project. The plan should be simple, so that updating is straightforward and does not demand the feedback of large amounts of data, and flexible, so that all alternative courses of action can be considered. This may be achieved either by allocating additional resources, or by introducing a greater element of float and extending the contract duration, as necessary. In either case, the estimated cost will increase, and hence it is essential to link the programme with the cost forecast.

It is difficult to enforce a plan which is conceived in isolation, and it is therefore essential to involve the people responsible for the constituent operations in the development of the plan. The plan must not impose excessive restraints on the other members of the organisation; it should provide a flexible framework within which they can exercise their own initiative. The plan must precipitate action and must therefore be available in advance of the task.

## 9.2  Programming

Programmes are required at various stages in the contract; when considering feasibility or sanction, at the pre-contract stage and during the contract. They may be used for initial budget control or for day-to-day implementation work. They may pertain to one contract, or a number of contracts in one large project.

The planner must therefore decide on the appropriate level of detail for the programme and the choice of programming technique. Important factors in this choice include the purpose of the programme, the relevant level of management and the level of detailed required. Simplicity and flexibility are the keys; a programme of 100 activities is easy to comprehend, whereas a programme of 1000 activities is not. Often it is good practice to ensure that the number of individual activities should relate closely to the basic packages of work required, or to the cost centres defined in the estimate, and should all have durations of a similar order of magnitude.

The period of time necessary to execute the work of an activity, the 'duration', depends on the level of resources allocated to the activity, the output of those resources and the quantity of work. The duration may also depend on other outside restraints, such as the specified completion date for the whole or some part of the work, the delivery date for specific material or restrictions on access to parts of the works.

A number of common forms of programme used in engineering project management are reviewed below.

### Bar charts

The most common form of plan is the bar chart, also known as the Gantt chart, and an example is shown in Figure 9.1. Each activity is shown in its scheduled position to give an efficient use of resources, the logic and float are shown by dotted lines, and important constraints or key dates are clearly marked. The space within the bars can be used for figures of output or plant costs, and there is room beneath to mark actual progress. It is frequently useful to plot a period-by-period histogram of the demand for key resources directly under the bars at the bottom edge of the programme.

### Line of balance

This simple technique was developed for house building, and is useful for any repetitive type of work. The axes are the number of completed units and time: the work of each gang appears as an inclined line, the inclination being related to the output of the gang, as shown in Figure 9.2.

### Location–time diagram

In cross-country jobs such as major roadworks, the erection of transmission lines or pipe-laying, the performance of individual activities will be greatly affected by their location and the various physical conditions encountered. Restricted access to the works, the relative positions of cuttings and embankments, sources of materials from quarries and temporary borrowpits, the need to provide temporary or permanent crossings for watercourses, roads and railways, and the nature of the ground will all influence the continuity of construction work and the output achieved by similar resources of men and machines working in different locations (Figure 9.3).

## 9.3 Network analysis

There are two basic forms of network analysis techniques: precedence diagrams, sometimes called activity-on-node networks, and arrow diagrams, sometimes called activity-on-line networks.

Although both methods will achieve the same answer, the precedence

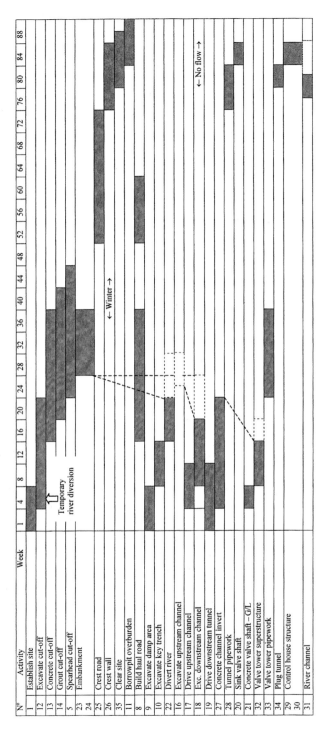

**Figure 9.1** Construction of a bridge: programme in bar chart form.

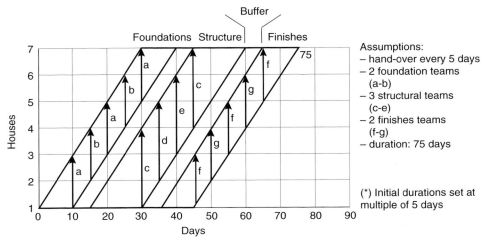

**Figure 9.2** Line of balance.

diagram is recommended by the authors. The advantages of the precedence system over arrow diagrams are:

❑ flexibility, since logic is defined in two stages;
❑ dummy activities are eliminated;
❑ revision and the introduction of new activities is simple;
❑ overlapping and delaying of activities is easily defined;
❑ use of pre-printed node sheets is possible.

The network diagram (Figure 9.4) resembles a flow chart, with activities being represented by egg-shaped nodes, and the interrelationships between activities by lines known as dependencies. The dependencies are developed by moving to each activity in turn, asking 'What can start once this activity has started?' and drawing the relevant lines. The convention is that dependencies run from nose to tail of succeeding activities, that is from the finish of one activity to the start of the succeeding activity. Figure 9.4 relates to the worked example below.

Apart from the 'critical path' a degree of choice exists in the timing of the other activities; a characteristic which is called 'float'. It is the float which will later be utilised to adjust the timing of activities in order to obtain the best possible use of resources. The 'total float' associated with an activity is the difference between its earliest and latest starts or finishes.

'Free float' is the minimum difference between the earliest finish time of that activity and the earliest start time of a succeeding activity. Total

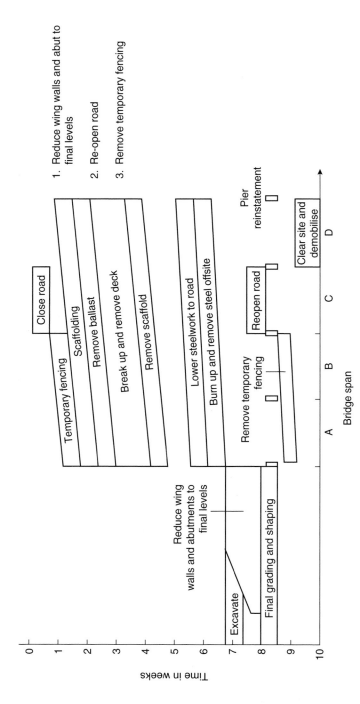

**Figure 9.3** Time location.

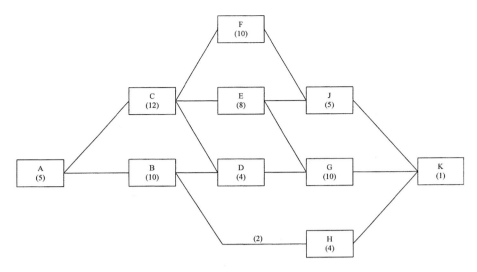

**Figure 9.4** Precedence diagram: worked example.

float is a measure of the maximum adjustment that may be made to the timing of an activity without extending the overall duration of the project: free float is that part of the total float which can be used without affecting subsequent activities.

To illustrate the techniques needed to perform the precedence diagram critical path method, a worked example based on a simple ten-activity network is described below.

**Worked example**

Table 9.1 shows the durations and interdependence of ten activities required to carry out a project.

(1)    Construct a precedence diagram assuming no resource restrictions and calculate the minimum duration of the project.
(2)    Schedule the earliest and latest start and finish for each activity and show the critical path.
(3)    If activity F is extended to a duration of 15 days, what is the effect on the critical path?

*Solution*
(1)    Start with the first activity, A, at the left-hand edge in the centre of the pre-printed sheet. Because it is a start activity it has only successor activities. Using the logic in Table 9.1, the start of activities B

**Table 9.1** Network example.

| Activity | Duration (days) | Logic |
|----------|-----------------|-------|
| | | Depends only on the completion of: |
| A | 5 | – |
| B | 10 | A |
| C | 12 | A |
| D | 4 | B and C |
| E | 8 | C |
| F | 10 | C |
| G | 10 | E and D |
| H | 4 | B (2 days) |
| J | 5 | E and F |
| K | 1 | H and G and J |

and C is dependent upon the completion of activity A, so these activities are plotted immediately adjacent to A. This process is repeated until activity K is plotted. The network should then be checked to ensure that there are no loops of activities, and that there are no unconnected strings of activities.

The next step is to complete the forward pass. The project will commence at the end of time-period zero. Hence the early start for activity A will be 0. The early finish is calculated by adding the activity duration to the early start, i.e. 0 + 5 = 5. The early finish of A is then the early start of both B and C, day 5, and the process is then repeated. Activity D has two predecessor early finishes from activities B and C; as D cannot start until both B and C are completed, the larger value for the early finish from C, day 17, becomes the early start for D.

The link between activity H and activity B is subject to an overlap of 2 days, i.e. the start of H can overlap the finish of B by 2 days. The early finish for B is day 15, and so the early start for H is day 13. The forward pass continues until activity K is reached, and an early finish for the project is calculated as 36 days.

(2)  The process is then reversed to calculate the backward pass. Using the same logic but starting at activity K, a late finish is determined. In this example, as in the majority of cases, the late finish is taken to be the same as the early finish, day 36. It should be noted that it is possible to insert a higher number as the late finish, but the backward pass would then not end at time period zero. By the same logic, the late start is calculated by subtracting the activity duration from the late finish. This process is repeated, taking the lower value

on the backward pass and adding overlaps until a late start for activity A is calculated. Unless this value is also zero, an error has been made.

Each activity will now have an earliest and latest start and an earliest and latest finish. A number of activities, including the start activity and the finish activity, will have identical early starts and late starts. The difference between the late start and the early start is known as float. The critical path or paths consists of the activities with zero float logically linked between the start activity and the finish activity. This gives a critical path of A–C–E–G–K.

(3)   Activity F is not critical, and it has a float of 3 days. By extending the duration by 5 days, the early finish of activity F is also extended by 5 days. The forward pass is continued to show that activity K now finishes at day 38, an extension of 2 days. When the backward pass is completed, it is shown that there is a new critical path and changed float durations for non-critical activities. The new critical path is A–C–F–J–K.

## 9.4  Updating the network

Updating should be done frequently to ensure that the network is relevant. The procedure is described below.

(1)   Identify and mark all activities on which work is currently proceeding as 'live' activities. This is important as it focuses attention on the future. What has happened up to the present date must be accepted, and concentration must given to replanning and scheduling future activities.

(2)   Introduce a new activity at the origin of the programme which has a duration equal to the time-interval between the start of the programme and the date of updating.

(3)   Change the durations of all completed activities to zero.

(4)   Calculate revised durations from the date of update for all live activities and all future activities, taking into account any changes in requirements or actual performance. Note that a completely new estimate of the amount of work remaining to be done should be made for each 'live or future' activity at each update. The revised activity duration is derived from this figure, and a reassessment value of the probable output of the relevant resources.

(5)   Evaluate the programme in the normal way.

## 9.5  Resource scheduling

In the initial stages of planning a project, a precedence network has been constructed. The data for this have usually been derived using the most efficient sizes of gangs and the normal number of machines, with the assumption that the resource demands for each activity can be met. Bar charts have then been drawn using earliest start dates and showing maximum float.

The next step is to consider the total demand for key resources. When considering the project or contract as a whole, there will be competition between activities for resources, and the demand may either exceed the planned availability or produce a fluctuating pattern for their use. This is known as 'resource aggregation'.

The next stage, usually known as 'resource levelling' or 'smoothing', utilises the project float. Float can be used to adjust the timing of activities so that the resource limits imposed and the earliest completion date are not exceeded. If the float available within the programme is not sufficient to adjust the activities, the planner could consider the resources given to each activity and assess whether the usage can be changed, thereby altering the individual activity durations. It is clear that the levelling of one resource will have an effect on the usage of others. In consequence, resource levelling is usually only applied to a few key resources.

In some cases it will still not be possible to satisfy both the restraints on resource availability and the previously calculated earliest completion data, and the duration of the project is then extended. The lower the limits placed on resources, the greater the extension of the completion period of the project when it is 'resource scheduled'. Once the key resources have been adjusted, a new completion date results. If this is not acceptable, the resource limits must be adjusted and the process repeated. When resource scheduling has produced a satisfactory solution, the start and finish dates for each activity are said to be their 'scheduled' values. It is probable that few scheduled activities will offer float.

## 9.6  Planning with uncertainty

The planner is often faced with a degree of uncertainty in the data used for planning. Most planning techniques inevitably use single-point, deterministic values for duration and cost, although in practice it is realised that there may be a range of values for these parameters.

There may be times when it is more appropriate to consider the uncertainties of the project. There are a number of ways in which this can be achieved, ranging from the simple techniques of using pessimistic, optimistic and most likely three-point values, through sensitivity analysis, to the more sophisticated Monte Carlo random sampling method. In all cases it causes the planner to adopt a more statistical approach to the data, and focuses attention on the uncertainties in the project.

When using such techniques, the planner is probably forced into using a computer. The mechanics and logistics of multi-value probabilistic models eliminate manual calculations on all but the simplest of construction programmes.

One planning technique not necessarily requiring such powerful computation tools is the decision-tree approach. Used as a means of judging between decisions, this technique can be of assistance in the early stages of planning. Full details of the technique can be found in many standard texts.

## 9.7 Software and modelling

The increasing commercial pressure to achieve predetermined time and cost targets, combined with the power of the desk-top computer, has led to a proliferation of project management software packages. These programs vary widely in terms of their modelling flexibility and simulation options, but all are designed to serve the same purpose, which is to provide project managers with the power to plan the time and cost out-turn in projects. All programs link time, cost and the resources of the project, and allow the project manager interactively to forecast the financial commitments for the project and to assess a range of scenarios to reflect likely change and uncertainty. This section will not consider individual commercial software products in detail, but will provide generic guidance on the selection and utilisation of software packages for project management.

In considering the purchase of project management software, the first criteria should be the intended use. Project management software ranges in cost from the hundreds to thousands of pounds, offering a wealth of features to cater for a varying range of project types and project management styles.

One end of the spectrum may be characterised by project management teams, managing multiple projects, with the need to prioritise and manage resources across projects, while collaborating in a team-based environment. At the other, is the stand-alone project manager, managing a single project, with fewer activities and resources. Hence the evaluation of the matrix of features and functions on offer in any given project management software should be done in the context of the perceived need and potential future requirements.

In drawing up a short list of software to evaluate, the operating system which the software must support should be established. The pervasive nature of the Microsoft operating systems may make this appear to be a moot point, but should you be among the new wave of LINUX users, or favour APPLE Macintosh computers, the availability of a given project management software package on your platform of choice could be a limiting factor.

The key features of a given software package may be grouped and evaluated under the headings of activity modelling, resource handling, analysis tools, integration options, usability and costs.

### *Activity modelling*

In general, project management software provides facilities for activity (task) definition, allowing projects to be modelled using network diagrams or Gantt charts. With network diagrams, a good package will offer automatic recalculation of the critical path, and activity durations based upon changes to resource allocation and other variables. In reviewing the activity modelling features, one should check for constraints on the number of activities per project, the links per activity, the types of activity, and even the characters in activity names (some packages allow comprehensive descriptions).

Some packages enable tasks/activities to be grouped logically in breakdown structures. These typically take two forms, work breakdown structures (WBS) that group activities under summary headings, and organisation breakdown structures (OBS) that allow responsibility for activities to be assigned to departments, functions or workgroups. If these features are of interest, check for constraints on the number of levels that project activities can be grouped under, as these can vary from as few as 10 to as many as 999. Also check whether the package offers graphical-tree editors to support the creation of breakdown structures. Other organisational features such as the use of sub-projects, and constraints thereof, should also be checked, particularly if it is intended to model large projects.

Support for customising project diagrams varies between packages, and filtering and layout options, formatting features and support for annotations should be evaluated against your needs.

### Resource handling

Project management software packages should support the definition of resources (material, equipment, human), and should allow you to quantify the resource, detail its availability and associate with it a cost or rate. Defined resources can then be assigned to activities, and the software determines resource usage per activity, and whether an activity is resource-constrained (trying to use more than is available). Any good package should be able to reschedule projects to overcome resource constraints, and provide resource levelling and smoothing routines to maximise resource use. It is important to understand how these features are implemented, as they vary between packages. For instance, some packages will allow 'efficiency factors' (coefficients to weight the effort required, based on the skill of the resource) to influence the scheduling outcome, others allow resources to be modelled as partially available, or as consumable.

Project management software typically uses calendars to define resource availability and usage; one should check that the facilities are flexible enough for one's needs. Does the package allow you to define your own working week, number of hours in the day, etc? How are holidays modelled? Can shift patterns be customised?

### Analysis tools

At the very least, a good project management software package should support probability analysis via PERT network diagrams, allowing the definition of optimistic, most likely and pessimistic durations for activities, so that the probability of completing the project within a given time-frame can be ascertained. Packages also offer 'what-if' analysis features that allow projects to be versioned, revised and compared with the original. Check that the maximum number of scenarios that can be saved meets requirements.

Many packages come with more sophisticated analysis tools, ranging from cost management functions, such as earned value calculations, cash flow forecasting and wastage analysis, to Monte Carlo method-based probability analysis. In considering such features, one should remember to review the implementation details, and ensure that the techniques used are compatible with one's requirements.

Reporting tools are also of great importance, as they allow you to generate project reports that focus on relevant aspects of the project. Ensure that the software supports the kind of reports you need. Also review the customisation options: good filtering, sorting and annotation facilities can vastly improve the quality of reports. A print preview option should be considered a must. Also check on the printing peripherals the package supports; the ability to print large network diagrams on large sheets using pen plotters may be attractive.

### Integration options

In today's world nothing is done in isolation. A project manager will employ not only project management software, but also e-mail packages, word processors, presentation tools, spreadsheets, personal information managers (PIMs) and the internet. Thus, in evaluating project management software, consider how the package will integrate into your existing environment. For instance, many packages allow the information in PIM calendars to be imported into resource calendars, and can exchange data between spreadsheets and databases. Others can publish reports in HTML, allowing easy publication on the internet. The ability to export charts and graphs directly to your presentation package will be appreciated when attempting to meet that looming briefing deadline.

For enterprise-based users, integration with accounting and enterprise resource planning (ERP) systems should also be considered. Some packages offer native interfaces to these systems, while others may require the use of a third-party tool.

The more sophisticated packages come with collaborative environments, which allow multiple users to work on the same project. All such environments offer a locking mechanism to prevent multiple users changing the same data at the same time. The level at which the mechanism locks data, i.e. project, task or resource, varies between packages, and should be checked against your needs. Also remember to consider how the collaborative environment will integrate with your current security policies, and ensure that only authorized users have access to appropriate project data.

### Usability

The advent of the graphical user interface (GUI) can make all tools look the same at first glance. However, in evaluating the ease with which a project management software package can be used, one may wish to consider some of the following criteria. Are menus intuitive? How easy is it to navigate through the menus? Can one quickly zero-in on the desired action? How easy is it to specify the properties of activities, resources, etc? Are there any wizards (automated guides) for common tasks? Can I customise menus to my needs? The easiest way to evaluate these criteria is by actually using the software; many vendors will provide evaluation versions (typically time-bombed to stop working after a set period of time) for this purpose.

It is important to evaluate the help facilities associated with the package. The presence of tutorials and 'How do I' sections can be useful. Inspect the user guides for clarity and depth. Also enquire as to the degree of technical support you can expect from the vendors. Some vendors offer manned help lines, others offer e-mail-based support. What is the process for reporting bugs and getting bugs fixed? Make sure you are aware of the potential time-lines; investigate the reliability of both the software and the vendor. Product reviews and user groups can easily be found on the internet.

Alongside all this, remember to check the performance characteristics of the software. For example, what minimum hardware specification (processor, hard disk space and memory) does it require? Does use of the software require an expensive, untimely upgrade of your current machines? Did you find the time to open projects, save projects, create activities, etc., acceptable?

## Costs

Beyond the analysis of features and functions, there is the question of cost. Quite simply, is the software package affordable. In considering costs, one should look not just at the sticker price of the software, but also the total cost of ownership. What are the service and maintenance costs? If training courses are required, what will these cost? Will new hardware need to be bought? What happens when new versions are released? Are you entitled to them, under your current purchase (sub-scription)? Or will you have to factor-in upgrade costs?

By addressing the total cost of ownership alongside the features required, the project manager will be well equipped to make a decision on which project management software to invest in.

## Further reading

Antill, J.M. and Woodhead, R.W. (1982) *Critical Path Methods in Construction*, 3rd edn, Wiley, London.

Chapman, C.B., Cooper, D.F. and Page, M.J. (1987) *Management for Engineers*, Wiley, London.

Smith, A.A., Hinton, E. and Lewis, R.W. (1983) *Civil Engineering Systems*, Wiley, London.

## Exercises

### 9.1 *New housing estate*

A new housing estate comprising four blocks of low-rise flats, a block of five-storey flats, and a shops and maisonettes complex. New water and sewerage systems are required, and an existing drain requires diverting. Attractive landscaping, fencing and screening will complete the site.

Produce a programme for the project, and resource-level the programme for (a) bricklayers, and (b) total labour.

*Restraints*

Target completion: 82 weeks.

Flats A and B to be completed as early as possible.

Shops and maisonettes not required until all other work on site is completed.

*Resource demand*

Flats, blocks A, B, C and D, per block:

| | |
|---|---|
| Excavation, foundation | 40 man-weeks |
| Brickwork | 120 man-weeks |
| Carpenters | 30 man-weeks |
| Finishing trades | 70 man-weeks |

Shops and maisonettes:

| | |
|---|---|
| Excavation, foundations | 120 man-weeks |
| Concrete frame | 150 man-weeks |
| Brickwork, blockwork | 150 man-weeks |
| Finishing trades | 250 man-weeks |

Five-storey block:

| | |
|---|---|
| Excavation, foundations | 190 man-weeks |
| Concrete frame | 300 man-weeks |
| Brickwork | 350 man-weeks |
| Finishing trades | 500 man-weeks |

| | |
|---|---|
| Site establishment | 4 men × 3 weeks |
| Drain diversion | 30 man-weeks |
| Completion of main drain | 24 man-weeks, 300-mm-diameter drain |
| Drain connections | 40 man-weeks, connections to main drain |

|                    |                                        |
|--------------------|----------------------------------------|
| Roads, 1st phase   | 60 man-weeks                           |
| Road surfacing     | 50 man-weeks                           |
| Paths              | 75 man-weeks                           |
| Fences and screens | 20 man-weeks                           |
| Water services     | 80 man-weeks, 150-mm-diameter ring main |
| Landscaping        | 100 man-weeks                          |

## 9.2 Pipeline

The contract comprises the laying of two lengths of large-diameter pipeline with flexible joints 16 km from pumping station A to reservoir B, and 8 km from B to an industrial consumer C. There is a continuous outcrop of rock from chainage 10 to chainage 13 km, and special river, rail or road crossings are to be constructed at chainages 2, 7, 13, 18 and 22 km.

Isolating valves are to be installed at the ends and at 3-km intervals along the entire length of the main, each providing a suitable flange and anchor for test purposes. Water for testing will be supplied free by the promoter following completion of the pumping station during week 33. The reservoir will be commissioned during week 35. The industrial plant will be completed by week 40, but cannot become operational until water is available. Pipes are available at a maximum rate of 1000 m per week (commencing week 1). The contractor is responsible for off-loading from suppliers' lorries, storing and stringing out. Each stringing gang can handle 1000 m per week.

Produce a time–location programme for the contract on the assumption that one stringing gang and four separate pipe-laying gangs are to be employed. The average rate of pipe-laying per gang may be taken as 300 and 75 m per week per gang in normal ground and rock, respectively. Testing and cleaning each 3-km length will take 2 weeks. A river, rail or road crossing is estimated to occupy a bridging gang for 4 weeks. Because of access problems, pipe-laying and stringing gangs should not be operating concurrently in the same 1-km length. Attention should be paid to manpower resources and continuity of work.

## 9.3 Industrial project

(a) Construct a precedence diagram as the master programme for an industrial project using the logic and estimated duration given in the table below. Determine the earliest completion date and mark the critical path.

| Activity | Duration (months) | Precedence activity |
|---|---|---|
| 1. Promoter's brief | 3 | — |
| 2. Feasibility study | 18 | 1 |
| 3. Promoter considers report | 12 | 2 |
| 4. Land purchase | 12 | 3 |
| 5. Site investigation | 4 | 3 |
| 6. Design stage I | 6 | 3 |
| 7. Design stage II | 4 | 5,6 |
| 8. Civil tender documents | 3 | 5,6 |
| 9. Specify mechanical plant | 3 | 6 |
| 10. Mechanical plant tender | 4 | 9 |
| 11. Civil tender | 2 | 7,8 |
| 12. Specify electrical plant | 3 | 9 |
| 13. Electrical plant tender | 5 | 12 |
| 14. Manufacture mechanical plant | 18 | 10 |
| 15. Design stage III | 4 | 7,10 |
| 16. Design stage IV | 4 | 13,15 |
| 17. Manufacture electrical plant | 20 | 13 |
| 18. Construction stage I | 6 | 4,11,15 |
| 19. Construction stage II | 12 | 16,18 |
| 20. Install plant | 6 | 14,16,17,18 |
| 21. Test and commission | 3 | 19,20 |

Estimate the float associated with activity 16.

(b)  The tenders for the electrical and mechanical plant have been received and are as follows:

| Plant | | Period of manufacture (months) | Cost |
|---|---|---|---|
| Electrical A | Activity 17 | 20 | £444 000 |
| Electrical B | | 16 | £600 000 |
| Mechanical C | Activity 14 | 18 | £600 000 |
| Mechanical D | | 16 | £700 000 |

If the promoter's profit is estimated to be £50 000 per month from the date of completion, which two tenders would you recommend?

### 9.4 Bridge

Using the precedence diagram method, produce networks for the construction of the bridge shown in Figure 9.1.

| Activity | Description | Duration (weeks) | Resource demand |
|---|---|---|---|
| 1 | Set up site | 1 | |
| 2 | Excavate left abutment | 6 | Excavation |
| 3 | Excavate left pier | 4 | Excavation |
| 4 | Excavate right pier | 4 | Excavation |
| 5 | Excavate right abutment | 6 | Excavation |
| 6 | Pile-driving to right pier | 7 | |
| 7 | Foundations for left abutment | 8 | Concrete |
| 8 | Foundations for left pier | 6 | Concrete |
| 9 | Foundations for right pier | 6 | Concrete |
| 10 | Foundations for right abutment | 8 | Concrete |
| 11 | Concrete left abutment | 9 | Concrete |
| 12 | Concrete left pier | 7 | Concrete |
| 13 | Concrete right pier | 7 | Concrete |
| 14 | Concrete right abutment | 9 | Concrete |
| 15 | Place beams for left span | 4 | Crane |
| 16 | Place beams for centre span | 4 | Crane |
| 17 | Place beams for right span | 4 | Crane |
| 18 | Clear site | 1 | Crane |

This exercise is in *three* parts.

(1)    Produce a network assuming that unlimited resources are available. Evaluate the network, showing the critical path and the earliest completion date.

(2)    By considering the network produced in (1), evaluate the effects of reducing the duration of pile-driving to 5 weeks.

(3)    As (1), but assume that the resources are heavily constrained. The total resources available are one excavation team, one concrete team for foundations, one concrete team for abutments and piers, and one crane team.

Suggestions for answers to these exercises can be found at the back of the book, before the index.

# Chapter 10

# Project Control Using Earned Value Techniques

The purpose of this chapter is first to describe project control theory in general, and then to explain and discuss the system of earned value that is now used by a number of organisations. The section on earned value covers all aspects of the approach, from definition to application, in a variety of situations.

## 10.1 Project control

The purpose of project control is to ensure that the project's status is reported in a consistent, cost-effective and timely manner to the project manager, so that any necessary actions can be taken and the status of the project reported to senior management. To do this, the project manager will need to hold regular meetings with the project team and have regular, meaningful reports provided in an efficient and timely manner. In addition, a process must be put in place to control change, including schedule controls, change requests, budget controls and re-plan. The model here has to be plan–do–review. The 'do' component could take any of the following forms:

- ❑ do nothing if the work is progressing as planned;
- ❑ refer on if the change lies outside the control of the project team;
- ❑ change the plan, although this has the inherent problem that future measurements will be against a new base-line, but it may be the preferred option if it can be clearly demonstrated that it will benefit the project;
- ❑ change the work, either the scope or the method of execution, e.g. change the quality, change the functionality or seek an alternative.

The role of the project manager is to:

❑  establish the system;
❑  allocate responsibilities;
❑  ensure that costs are properly allocated;
❑  ensure that payments are authorised as appropriate.

This can be monitored by:

❑  obtaining regular, verbal reports;
❑  progress reports;
❑  internal measurement procedures;
❑  external audits.

This has to be balanced by the needs of the project team. Although they do need feedback on how they are performing, they also need to be clear about the need for the information that they are providing (that it is not just more bureaucracy) so that they do not feel constrained by the reporting system. If the focus is on reporting by exception, when there is a clear deviation from what is expected, then 'information overload' can be avoided and the reports should be timely.

In 1967, the United States Department of Defence published a set of cost/schedule control system criteria, known as the C-Spec. These criteria define minimum earned value management control system requirements. Previous management control systems assumed a direct relationship between lapsed time, work performed and costs incurred. This chapter describes how the earned value system analyses each of these components independently, and compares actual data with a baseline plan set at the beginning of the project. First, however, there will be a brief description of the *planning and control cycle* (Figure 10.1) which forms the basis of the methodology for controlling project costs.

The cycle consists of monitoring actual performance against a plan, using feedback to update and revise the plan on a regular basis.

## 10.2  Earned value definitions

*Earned value analysis* (EVA) compares the value of work done with the value of work that should have been done.

*Budgeted cost of work scheduled* (BCWS) is the value of work that should have been done at a given point in time. This takes the work planned to have been done and the budget for each task, and indicates the portion of the budget planned to have been used.

*Budgeted cost of work performed* (BCWP) is the value of the work done at a given point in time. This takes the work that has been done and the

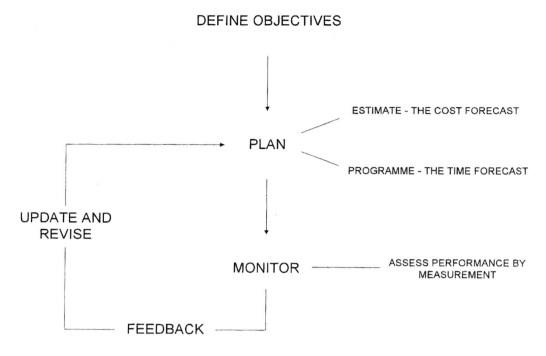

**Figure 10.1** Planning and control cycle.

budget for each task, and indicates what portion of the budget ought to have been used to achieve it.

*Actual cost of work performed* (ACWP) is the actual cost of the work done. Figure 10.2 illustrates a typical S-curve plot comparing budget and actual costs.

*Productivity factor* is the ratio of the estimated man-hours to the actual man-hours.

*Schedule variance* (SV) is the value of the work done minus the value of the work that should have been done (BCWP − BCWS). A negative number implies that the work is behind schedule.

*Cost variance* (CV) is the budgeted cost of the work done to date minus the actual cost of the work done to date (BCWP − ACWP). A negative number implies a current budget overrun.

*Variance at completion* is the budget (baseline) at completion (BAC) minus the estimate at completion. A negative value implies that the project is over budget.

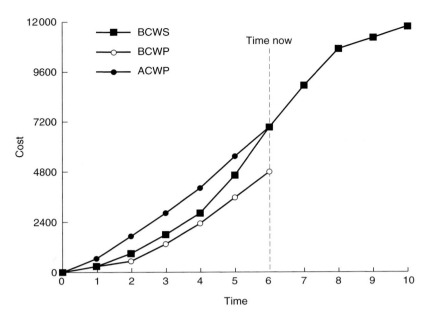

**Figure 10.2**  Typical S-curve.

*Schedule performance index* (SPI) is (BCWP/BCWS) × 100. Values under 100 indicate that the project is over budget or behind schedule.
*Cost performance index* (CPI) is (BCWP/ACWP) × 100. Values under 100 indicate that the project is over budget.

## 10.3  The theory and development of earned value analysis

S-curves examine the progress of the project and forecast expenditure in terms of man-hours or money. The result is compared with the actual expenditure as the project progresses, or the value of the work done. If percentages are used, the development of useful data from historic records of past projects is simplified, since size, and hence total time and cost, is not significant. All projects, whatever their size, are plotted against the same parameters, and then characteristic curves can more readily be seen.

The form of the S-curve is determined by the start date, the end date and the manner in which the value of the work done is assessed. Once a consistent approach has been established and the historical data analysed, there are three significant variables that need analysing: time, money and the shape of the S (known as the route). Since the expectation is that the route is fixed, then only two variables are left. The route is as much a target as the final cost. If the movement from month to month is compared, then the trend can be derived.

If the axes of the S-curve are expressed as percentages, then what the percentages are based on must be carefully defined. If review estimates are made at regular intervals, then the value of the work done will always be a percentage of the latest estimate. Revision of the estimate automatically revises the route of the S-curve.

The assessment and precise recording of the value of the work done is crucial to project cost control. This is described by accountants as 'work in progress'. For example, design man-hours are usually measured weekly. The hours that have been booked can be evaluated at an average rate per man-hour, that rate being the actual costs for the project. Materials are delivered against a firm order, so normally an order value is available. The establishment of realistic targets is very important if the analysis is to be meaningful. A low cost estimate often leads to a low estimate of the time required to carry out the work. The immediate target is not the final target.

A series of standard S-curves has been developed by companies in the oil and gas industries so that the performance of existing projects can be monitored. These curves have been derived empirically, and when projects within certain categories do not follow the norm, investigations ensue to identify the source of the discrepancy. These curves have been put to a number of uses, including monitoring, reporting and payment. They are not always used in the pure form, i.e. further curves, such as productivity factors, can be developed so that certain aspects of the project can form the focus of attention.

Developments have been made from using the S-curve simply as a method for controlling the progress of the project to using it proactively in the evaluation of indirect costs associated with project changes introduced by the promoter. This method is known as 'impact or influence' (Figure 10.3).

Influence may be applied whenever there is a variation, and it is applied to the estimate of additional man-hours by taking into account the indirect costs for the whole of the variation, including the additional indirect costs associated with parallel activities. The revised estimated cost of the variation may then be issued to the promoter.

The influence is composed of:

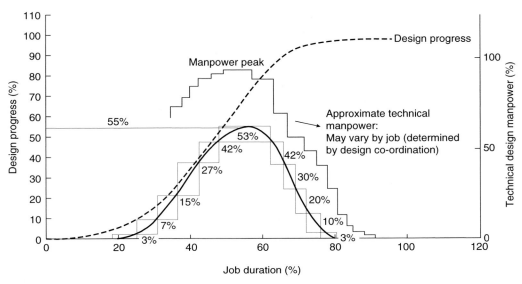

**Figure 10.3** Influence curve.

❏ time lost in stopping and starting current activities in order to make the change;
❏ special handling to meet a previously scheduled activity;
❏ revisions to project reports and documents;
❏ unusual circumstances which could not be foreseen;
❏ recycling (lost effort on work already produced);
❏ other costs which may not appear to be related to a particular change.

Influence is incorporated by multiplying the direct cost for the variation by the influence factor and adding this to the direct cost to give the total cost for the change.

$$T = V(1 + IF)$$

where $V$ is the direct cost of variation, $T$ is the total cost of variation, $IF$ is the influence factor and cost is the cost to the promoter. Any organisation wishing to adopt this approach would need to derive standard curves.

## 10.4 Relationship of project functions and earned value

Monitoring and cost control can be described as identifying what is happening and responding to it. Cost planning involves forecasting how money will be spent on a project in order to determine whether the project should be sanctioned, and having sufficient money available when required.

Typically, there are three major areas of control: commitments, value of work done and expenditure. These are all controlled in relation to their progress over time and may be illustrated diagrammatically, as shown in Figure 10.4, which presents a typical S-curve for the value of the work done. Similar S-curves may be developed for both commitments and expenditure.

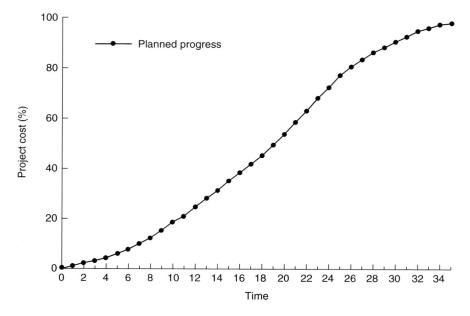

**Figure 10.4** S-curve showing the value of the work done.

Planning is primarily responsible for establishing the time target, both overall and in detail. The primary, although not the only, task of project cost control is to establish the exact position of the project from one time period to another in terms of the value of the work done, and compare this with the targets for each time period. Finance will maintain expenditure records, working closely with project cost control. Finance will also be responsible for maintaining a record of commitments, since they have to ensure that payments are within the approved limits.

Project control can comment on the validity of planning work by comparing planned and actual progress via the value of the work done. The commitment record, if properly maintained, provides a ceiling at any time during the project's life.

## 10.5  Value of work done control

The value of the work done is not the expenditure, although it eventually equates to expenditure. It is often considered as work in progress, but most of the time it is greater than the expenditure. Figure 10.5 illustrates the relationship between value of the work done and the expenditure as recorded over the life of a project.

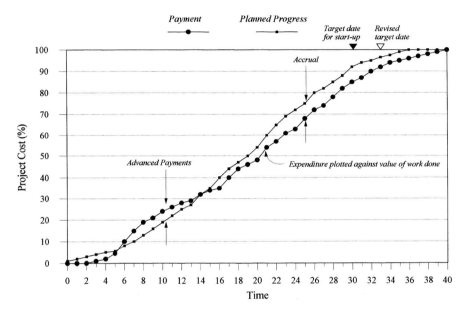

**Figure 10.5**  Comparison of work done and expenditure.

There is normally a considerable time-lag between having a project ready for start-up and the final payment of invoices and retention when the project is completed in financial terms. The value of the work done can be summarised as design and head office costs plus the value of the material delivered to the site and the work done at the site.

The techniques for reaching an approximation of the value of the work done from month to month can be related to three major areas: head office, materials delivered to the site and erection.

### *Head office*

Work done at head office includes design, procurement and project supervision activities. The work done can be measured in man-hours, and is usually recorded on a weekly basis. Hours booked are valued at an average man-hour rate, and the value of the work done is then assessed. The simple way to do this is on a cumulative basis. If the latest cumulative booked cost comes perhaps 3 weeks later, then that is compared with the cumulative cost including the last month's estimate. A correction can then be made, resulting in the approximation being only one month behind the actual cost at that time.

### Materials delivered to the site

Materials are normally delivered to the site against firm orders so that an order value is available. As materials arrive on-site the materials receipt note can be valued, using the information on the order, and a progressive total can be maintained. Since the value shown on the order may not be the absolute invoice value, for reasons such as discounts, freight charges, insurance or escalation, this approach gives a nominal and close approximation of the value.

In the case of bulk materials, it can be very time consuming to value all items on a receipt note from the itemised prices on an order, and it is often permissible to use a weighbridge weight multiplied by an average price per kilo calculated from the order or orders.

### Erection

The approach to erection contracts is very similar, but this time the value is erection man-hours. Erection man-hours, exclusive of site supervision, for all contracts on site, should be recorded from week to week. Progress on site by various erection contractors is measured by teams of schedulers or measurement engineers in accordance with certain standard methods of measurement to allow progress payments to be made. Such evaluations often run late so erection man-hours worked are often valued on the basis of standards multiplied by the average rate per hour which is applicable to the erection activity.

## 10.6  Earned value analysis techniques

Earned value analysis (EVA) is often presented in the form of progress, productivity or S-curve diagrams. Many of the current reporting procedures were developed by the US Corps of Engineers. EVA has been adopted by the oil, manufacturing, gas and process industries, where man-hours and material deliveries to the site are used to monitor projects from inception to completion.

Actual/estimated man-hours are made available to determine the progress and productivity factors at any stage of the project. The productivity factors are used by both promoter and contractor organisations to monitor the progress of a project and forecast the outcome.

Productivity is the ratio between output and input, and provides a measure of efficiency. Ideally, productivity is always unity, with both the output and the input measured in the same units. In the oil and gas industries, man-hours are used as the common unit to determine productivity. Productivity factors are used to monitor variance and trends for individual activities.

In order to establish a trend actual progress must be measured and compared with forecast progress, that being the base-line S-curve. Since the forecast progress depends upon the end target, revised targets will influence the progress which must be made each month. Once it becomes evident that a work package is going to cost more, or less, than the original (or earlier) estimate, then targets should be revised, the potential influence on monthly progress evaluated and a new target computed.

Table 10.1 illustrates a typical 'progress of value of work done' chart, expressed in both man-hours and completed items, which is common to many manufacturing industries. The revised number of man-hours is forecast against the actual work completed, and is based on progress and productivity. This information often is displayed on a screen to manufacturing operatives as a guide to production rates and the time to completion. If a bottleneck occurs at any point in the production line the affect on the costs, time and productivity of the completed item can be identified. In cases where production is carried out at a number of locations with final assembly at one specific point, it is imperative that any deviation from the activity S-curve is reported immediately and acted upon.

**Table 10.1** Progress of value of work done.

|  | Man-hours | Units |
| --- | --- | --- |
| To date |  |  |
|    Estimated | 360 000 | 4750 |
|    Actual | 295 000 | 4900 |
| Total |  |  |
|    Estimated | 500 000 | 6000 |
|    Revised | 440 000 | 6000 |

In many manufacturing industries, especially automobile assembly lines, man-hours are allocated for re-working items to meet quality and standards. Mercedes Benz, for example, reportedly spend up to 30% of the total man-hours allotted to re-working to meet defined quality and standards. This 'man-hour float' is often the determinant for price setting to the customer for each model. The hours expended from the 'float' of man-hours is then used to forecast the final costs, times and margins. In many cases, re-working is analogous to commissioning, and as such, may not be allotted targeted man-hours in the project estimate.

## 10.7  Application of EVA

The information required from an analysis of the curves will vary depending on its end use. At project level, the aim will be to identify any areas where the project is under-achieving in order that action can be taken to improve the performance of the problematic resource. This can be done by examining the BCWP and ACWP curves. If the curves are showing cause for concern, then productivity curves can be plotted and examined in greater detail. Work packages, or sets of work packages, can be examined and the productivity can be derived.

A project manager can adopt this system through the analysis of completed time sheets. It is also important to note the percentage-complete status of a task at the recording date, as described earlier. This is where a proactive role is required by the project management team in gleaning information from the various disciplines. The team leader should ensure the progress of the work of his section, with the planners taking a recording and reporting function. If progress is to be expedited, then the project team must actively pursue information, check its accuracy and take action when low productivity becomes apparent.

Optimum workforce requirements will become apparent as data for specific projects are collated. Productivity increases or decreases will permit management decisions based on actual measurements, rather than optimistic or pessimistic forecasts based on 'gut feeling'. The team leaders of various disciplines must be aware of the significance of the information they are reporting and why it is being recorded; if they do not realise that it will be used for control and not just for monitoring, they may not give it a high priority, resulting in erroneous data.

Cost codes for both direct and indirect costs will need to be considered on the basis of compatible codes allocated to different work packages. In a number of major process organisations in the oil and gas industries, for example, the number of multi-disciplinary functions and development stages of a project have led to confusion about the cost codes, resulting in inaccurate data and delays in transmitting data for analysis. Ideal cost-coding systems are those that allow the system to be used at all stages in the development of a project.

One very important aspect of coding and recording man-hours is the comparison of man-hours expended against a code or codes and the remaining man-hours expended as overheads. If, for example, a design engineer is allocated 40 h per week on, say, three projects and only 32 h are recorded, then the remaining 8 h must be reconciled somewhere within the organisation. Unfortunately, the remaining man-hours are often distributed between the project codes. In cost-plus contracts, it is standard procedure to book additional man-hours against the contract code.

To achieve the introduction of EVA, a basic requirement is the collection and processing of data related to existing and past projects. The importance of developing standard curves cannot be over-emphasised. If there are no standard curves, then it is difficult for the project manager to set realistic targets.

These data form the basis for estimating and allotting man-hours to each activity or work package. The effort required at project level to undertake EVA is such that it is not recommended for 'small' projects. Below a certain size, the effort required to gather and process data on the progress of small-value items may be greater than the value of the package of work. The proportion of the inaccurate part of the assessment of the work left will also be greater, thus invalidating the approach.

If the decision is made to adopt EVA for a given project, then the management team must be fully dedicated to implementing the system. Taking action on inaccurate data can be worse than taking no action at all. A standard reporting technique should be developed for all projects so that a given productivity rate means something.

## 10.8   Examples of EVA

Table 10.2 illustrates the historical data of a typical six-package project. Estimated man-hours are often based on historical data from completed projects, against which the actual man-hours expended to complete each work package and the overall project are plotted.

Figure 10.6 illustrates the S-curve for the prototype work package shown in Table 10.2. The S-curve is prepared on the basis of estimated man-hours. The actual man-hours expended are then plotted on a regular basis to determine the variance in man-hours and the productivity factors, and to forecast the work package trend.

**Table 10.2** Work packages, man-hours and productivity factors.

| Work package | Actual man-hours | Estimated man-hours | Productivity factor |
|---|---|---|---|
| Feasibility study | 550 | 450 | 0.82 |
| Design | 3000 | 2500 | 0.83 |
| Prototype | 1500 | 1700 | 1.13 |
| Manufacture | 2300 | 2200 | 0.96 |
| Erection | 900 | 600 | 0.66 |
| Commission | 300 | 300 | 1.00 |

On a number of civil engineering projects, EVA has been used primarily to monitor the construction of individual units and as a reporting tool to both construction management and the promoter. In its simplest form, bar charts are prepared from which histograms of man-hour estimates for each construction activity are generated. The cumulative man-hours are plotted in the form of an S-curve and progress is monitored as man-hours are expended. Plant usage is often added to man-hour schedules based on weekly costs, usually plant hire rates, for specific types of plant/activity usage. The type of plant and estimated

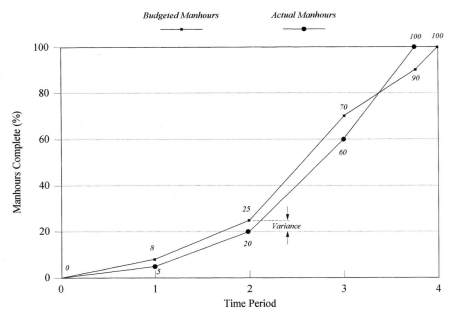

**Figure 10.6** S-curve for prototype work package.

usage often form the basis of plant allocation for a number of projects. Clearly, if the productivity of one particular work package falls below the estimate, as illustrated by the S-curve, then additional plant time and usage will be required to reverse the trend.

The diagrammatic presentation facilitates the comparison of actual man-hours expended in construction with the man-hours estimated. If, for example, in a fixed-price construction project the foundation works of a particular unit has a programme duration of 4 weeks utilising 480 h and is completed in 3 weeks utilising only 320 h, then the saving in man-hours can be utilised on another construction unit. The saving in man-hours and plant usage and the reduction in duration can then be plotted as actuals against the estimated S-curve. The 'productivity factor' for this completed activity would be 1.5.

If, however, 480 man-hours have been expended after 4 weeks with only 50% of the activity completed, then completion would require another 4 weeks and an additional 480 h based on the previous productivity factor. This often results in man-power being re-allocated to an activity which is falling below the estimated productivity factor. In addition, the requirement of plant to complete the work package will often result in delays to other work packages, as it is normally very difficult to acquire plant at short notice.

A major problem area is the accuracy of reports of percentages of work completed. As a result, activities or work packages are often reported as being completed when in fact they are not. This is often referred to as the 'persistent 99% complete syndrome', which results in the saying that 99% of tasks in 99% of projects are 99% complete for 99% of the time. It should also be noted that the method of appraising work completed should be timely and consistent for each section of work, and always kept up-to-date. In many civil engineering contracts, managers prefer to rely on the 'remaining duration' as a good measure of progress for a work package, as this does not assume anything about the accuracy of the estimate or when the work package started. This approach also ensures that the manager is aware of the remaining duration to meet any contractual milestones, and will act accordingly.

EVA is also used in the manufacturing and production industries to monitor production rates. In these industries, work is normally performed in controlled environments where it is a simple task to measure the physical work done and the materials used, especially where the delivery of an item is paramount to the success of a project. By monitoring, often on a daily basis, the number of man-hours both estimated and expended on a number of activities, delivery dates can be identified at an early stage of the project. The cumulative man-hours expended for each activity are compared with the estimated target of man-hours and the completed items, which permits adjustments to the overall target over the manufacturing period. Changes in the base rates per hour for each activity can easily be amended to identify the cost to completion. In production-line industries man-hours expended can be collated on a daily basis through timesheets, and completion can be verified and hence accurately reported on the basis of a completed item.

Many of the US companies involved in the oil and gas industries in the Gulf of Mexico introduced EVA to major projects carried out in the Middle East. In most cases, these projects were undertaken on a cost-plus basis. Data in the form of man-hours relating to specific tasks were analysed, and then utilised to provide 'productivity factors' relating to work carried out on projects in the US. To compensate for the time required to learn specific tasks, productivity were weighted to take account of this additional time.

For example, the work package for the cold insulation of pipe-work may have a US base-rate of 1 man-hour to insulate 1 m of 200 mm-diameter pipe. To take account of the possible learning period for this activity, which is often performed by third country nationals (TCN), a factor of, say, 1.2 is used. The S-curve is then factored by 1.2, and this factor is called the US Gulf factor (USGF). If the total number of estimated man-hours is, say, 200 000 h, then the factored total will be 240 000 h.

The man-hours can then be scheduled for this work package against time, as illustrated in Table 10.3. Clearly, the overall 'productivity factor' for this work package is 0.96, but productivity over the 9-month dura-

**Table 10.3** Estimated/actual man-hours and productivity factors.

| Month | Estimated man-hours | Actual man-hours | Productivity factor |
|-------|---------------------|------------------|---------------------|
| 1 | 10 000 | 18 000 | 0.55 |
| 2 | 40 000 | 50 000 | 0.80 |
| 3 | 60 000 | 70 000 | 0.86 |
| 4 | 50 000 | 48 000 | 1.04 |
| 5 | 40 000 | 30 000 | 1.33 |
| 6 | 20 000 | 15 000 | 1.33 |
| 7 | 10 000 | 8 000 | 1.25 |
| 8 | 9 000 | 10 000 | 0.90 |
| 9 | 1 000 | 2 000 | 0.50 |
|  | 240 000 | 251 000 | 0.96 |

tion has varied from 1.33 to 0.5. This method of reporting and planning at timely intervals provides management with a clear view of progress and productivity, and provides the basis for management decisions. In this particular example of a cost-plus project, the additional man-hours expended to meet the contract programme are paid for by the promoter.

It is very important that work packages utilised in EVA are defined correctly in the disciplines to be adopted, the allocation of man-hours and the inter-dependency on other work packages. The overall project is determined by the worst work package. In a project consisting of, say, 17 work packages, the productivity of 16 of the packages may be above that estimated. In a number of projects, contractors have superimposed individual S-curves to provide a total contract S-curve, which is normally the basis for reporting to the promoter. This often results in a false picture of the project's progress, since the total contract S-curve implies that the project is ahead of schedule.

In one example, schedulers on one particular work package reported that productivity was as estimated, but the man-hours recorded against the actual work completed were optimistic. This resulted in the project being reported as ahead of time and below budget. When an audit of the work package was performed, it was found that unless this critical work package was brought back on programme, the remaining 16 work packages could not be completed.

An additional 30 000 man-hours had to be allotted to the work package, which resulted in an overall productivity factor of 0.33. Clearly, the accuracy of reporting actual work and progress is a prime function of EVA. If effective reporting and updating had been carried out, then at some time during the activity the negative progress would have alerted management to the apparent problems.

A further problem on this particular project was monitoring the receipt of materials delivered to the site. The numerous cost codes, often with more than one cost code being applied to pre-assembled units, resulted in delays in processing invoices. In some cases a 3-month lag occurred between the receipt of materials and the processing of invoices.

To ensure that the same mistakes did not occur again, the promoter instructed the contractor to change the scales on each work package S-curve. Clearly, when a project has an estimated man-hour expenditure of 11 000 000 h, a slight deviation on the graph does not fully represent the impact on the project as a whole. Over the remaining project duration, each work package was expressed in total histogram form with 'productivity factors', time to complete and trend analysis being illustrated for presentation to the promoter.

*Project reporting*

The control reports should facilitate:

❏  agreement of customer requirements;
❏  optimisation of data requirements;
❏  standard formats;
❏  minimal distribution lists;
❏  common understanding of terms;
❏  paperless systems, where possible.

The project manager should agree on the reporting requirements as early as possible in the project's life. These requirements will be dependent upon size, complexity and customer requirements, and need to cover:

❏  level of detail;
❏  recipients;
❏  schedule;
❏  level of reporting.

Wherever possible, data should be collected from existing systems. It is sensible to format reports so that details can be summarised as necessary for upward transmission.

## 10.9  Summary

Standard procedures need to be prepared as the EVA system is tested and the requirement of each discipline is addressed. As with many reporting and scheduling systems, it is essential that the organisation uses and develops the system to suit their own needs, and not to create mountains of irrelevant information.

EVA is primarily a system of approximation, the accuracy of which depends on the time and costs prepared in the estimate compared with the actual time and costs as work progresses. The accuracy of the estimated and actual data and the time intervals for auditing are paramount to its successful application.

## Further reading

Kharbanda, O.P. and Stallworthy, E.A. (1991) *Cost Control*, Institution of Chemical Engineers, Rugby.

Maylor, H. (1999) *Project Management*, 2nd edn, Financial Times, Pitman, London.

Reiss, G. (1996) *Project Management Demystified*, Spon, London.

Turner, J.R. (1997) *The Commercial Project Manager*, McGraw Hill, London.

# Chapter 11

# Contract Strategy and the Contractor Selection Process

Contracts are used to procure people, plant equipment, materials and services, and consequently are fundamental to project management. This chapter outlines the main components in the selection process for both contract and contractor. The processes are influenced by the nature of the parties included, the project objectives, and the equitable allocation of responsibilities and risk, amongst other factors. This chapter is not meant to be a substitute for the expertise and thinking needed in drafting and administering any individual contract. It is an introduction to the relationship between legal means and engineering ends.

## 11.1 Context

Every project involves starting from one point and getting to a different point. In other words, the project is always about achieving a result. The problem for the promoter of the project is that he cannot, or does not wish to, provide all the resources necessary to get from the starting point to the final result from internal sources. Therefore there is a need to obtain resources, work, materials, equipment or services from other organisations in order to achieve the desired result. The method by which the promoter will obtain those resources is that of the contract. A proper understanding of how contracts work and how they must be managed is therefore fundamental to the management of virtually all projects.

In addition, it has to be recognised that different industries use different types of contract to achieve different results. Indeed, there is an almost complete cultural divide between some industries. The standard type of activity/work-based contract used within the building and construction industries is totally different to the result-based contracts used within the equipment process or oil industries, and within service industries such as the telecommunications or the software industry.

In addition, most complex projects can be treated in several different ways. They can be carried out under a single 'turnkey' contract. They can be broken down into separate contracts on a time/stage basis, with the preliminary design being carried out under one contract, site preparation under a second contract, equipment manufacture under a third contract, and so on. They can be broken down into separate contracts on the basis of contractor skills, with one specialist contractor being responsible for all civil engineering and site preparation work, another being responsible for the supply of the main plant and equipment, a third being responsible for the supply of ancillary equipment, and a fourth being responsible for installation/erection. They can be broken down into a large number of small contracts or a small number of large contracts. Each will give the promoter different advantages and disadvantages. The promoter must decide.

Finally, different contracts can operate in different ways. In practice, the way in which a contract operates will generally depend to a large extent upon the way in which it provides for payment by the promoter to the contractor. In very broad general terms, contracts provide for payment in three different ways. The first is that of the price-based contract, under which the contractor will provide the bulk of the work materials/ equipment or services for a stated price or fee. At the opposite end of the scale is the reimbursable contract, under which the contractor is reimbursed with the costs of carrying out the work plus a profit. Finally, somewhere in between these two, is the quantities or rates-based contract, such as the typical civil engineering/building contract, in which payment is based upon the prices stated in a bill of quantities (BOQ) or a schedule of rates. The different types of contract tend to create different relationships between promoter and contractor, and therefore to produce totally different results. The type(s)/size(s) of contract should be selected by the promoter only after due consideration of the nature of the parties to the project, the contract management resources available, the project objectives and the skills required to achieve them, the time available to carry out the project, and the appropriate allocation of duties, responsibilities and risks.

## 11.2 Factors affecting strategy

In deciding the choice of contract strategy, a number of factors should be considered. Clear definitions of the promoter's objectives are required so that the significance of these factors can be established.

The responsibilities of the parties need to be determined. Responsibilities may include design, quality assurance and control, operating decisions, safety studies, approvals, scheduling, procurement, construction, equipment installation, inspection, testing and commissioning. The risks must then be allocated between the parties; in other words, who is to bear the risks of defining the project, specifying performance, design, selecting sub-contractors, site productivity, mistakes and many more. Then the basis for payments for design, equipment, construction and services must be decided. These are all major influences on the choice of contract strategy.

It is also important to consider the provision of an adequate incentive for efficient performance from the contractor. This must be reflected by an incentive for the promoter to provide appropriate information and support in a timely manner.

The contract may need to be flexible. The prime aim is to provide the promoter with sufficient flexibility to introduce change that can be anticipated, but is not defined, at the tender stage. An important requirement is that the contract should provide for systematic and equitable methods of introducing management and pricing changes. All these considerations are closely interrelated.

It is apparent that generally the interests of the promoter and the contractor tend to be opposed to each other. For example, a lump-sum contract minimises the cost/price risk carried by the promoter, but it imposes maximum cost/price risk on the contractor. The converse is true at the other extreme of a cost-reimbursable plus percentage fee contract.

Many, or almost all, of these factors could be important for any one project, but it is likely that certain factors will dominate. All factors are significant, but they may also conflict. If the work is on an off-shore oil energy project, time-scale might be the dominant objective in order to meet a weather window or to avoid disruption of the flow of crude petroleum; this would have obvious implications for the cost of the project. If working on a project for a plant to manufacture pharmaceuticals, the quality of plant and its ability to manufacture the correct product might be dominant.

There will often be a number of possible strategies which could satisfy these objectives, and it is the task of the project manager to advise the promoter which strategy to adopt. This selection is possibly one of the most important decisions in any project.

There can be other criteria to consider, and every project has to be assessed individually. For example, changes and innovations in contract arrangements have followed the privatization of what were formerly public services in the UK, to try to meet commercial rather than political accountability.

## 11.3  Contractual considerations

First, the project manager should always remember the three 'Rs' of contract management (Figure 11.1).

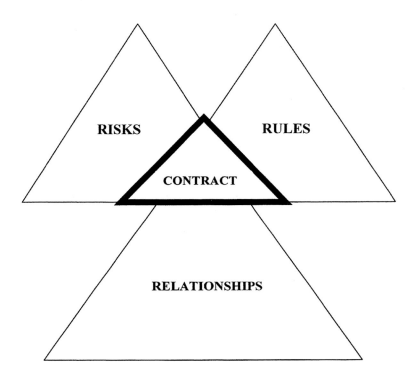

**Figure 11.1**  The three Rs.

(1)  *Relationships.* Relationships form part of all commercial dealings. Some relationships are permanent, some are temporary. Some relationships make for collaboration between the parties. Some relationships make for arm's-length dealings. Some are highly structured. Some are unstructured and will develop according to the way that circumstances dictate. Different types of relationship need to be managed in different ways. A collaborative relationship requires a very different management style to an adversarial relationship.

(2) *Risks.* Risks are present in all businesses and transactions. Risk does not go away, it is always there in one form or another. Whatever contract strategy the promoter chooses, it will bring with it its own particular set of risks for the promoter and the contractor. It is always important for the promoter to select a strategy that produces a risk set that he is able to manage competently. (It is also important to select a contractor who is also competent to manage that risk set. A contractor who can carry out work competently, but who cannot manage risk properly, is potentially disastrous.) Every risk has to be managed by one or other of the parties involved. If risk is not managed, and something goes wrong, then the problem will be much more serious.

(3) *Rules.* Rules are created in order to regularise the way in which organisations and individuals work in relation to each other. Comply with the rules and you have a reasonable chance of being successful. Fail to comply with the rules and, however brilliant you are, you will probably fail to achieve success unless the other party is prepared to be forgiving.

Contracts are one area where the three Rs overlap each other. Contracts formalise a set of risks, rules and relationships into one set of words which will govern all dealings between the parties while carrying out that contract.

Second, the promoter needs to remember the constraints imposed by the law. This is obviously not the place to discuss the law of contract in any detail. However, it is very important to understand that the law, whatever the country, takes a very artificial and simplistic view of the contract relationship. This view is best explained by using two similes.

(1) Making a contract is like jumping off a cliff. It is entirely up to the parties to decide whether or not they wish to enter into the contract. However, once they have made the decision and entered into the contract, then there is no going back. They have to live with the deal that they have made, as set out in the contract, and carry out the terms of that contract, unless both parties agree not to do so. If one party gets itself into the 'wrong' contract there is no easy option.

(2) The contract is an egg. Once an egg has been laid there can be no direct contact between what is outside the eggshell and what is inside the eggshell until the egg hatches. The egg is its own little world, with no relationship to anything else during that time. The law sees a contract in exactly the same way. The written contract is seen as a complete and precise statement of all the provisions

agreed between the parties relating to that contract. Nothing else can apply, and what goes on inside the contract is entirely divorced from any other relationships that the parties may have between them.

## 11.4 Contractor choice

'Normal' contract objectives are usually said to be cost, time and quality. They are often pictured as a triangle (Figure 11.2).

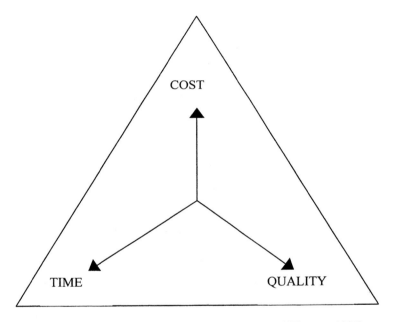

**Figure 11.2** Project objectives (adapted from Barnes and Wearne, 1993).

The use of the triangle is intended to illustrate the conflict between these three objectives. If the promoter wants the lowest possible cost, then it will not be in the shortest time-scale, and nor of the highest quality. If it is the best quality, the time-scale will not be short and the cost will be high. If the shortest possible time-scale is wanted, then similarly quality may suffer and cost will also have to be sacrificed. These statements are really so obvious as to be virtual truisms. However, they do illustrate the basic problem of contract strategy. Everything has to be a compromise between conflicting objectives.

In the planning of a contract, the promoter needs to consider carefully the reason for employing a contractor. A promoter generally selects a contractor for one or more of the following reasons.

❑ To use the particular management, technical and organisational skills, and expertise of that contractor for the duration of the project.
❑ To use the skills of the contractor after the project has been completed.
❑ To have the benefit of the contractor's special resources, such as licensed processes, unique design or manufacturing capability, plant, materials in stock, etc.
❑ To get work started quicker than would be possible by recruiting and training direct employees.
❑ To get the contractor to take some of the cost risks of a project, usually the risks of planning the economical use of people, plant, materials and sub-contractors.
❑ To use the contractor to provide the resources, both physical and financial, needed for the project.
❑ To be free to use his own (limited) resources for other purposes.
❑ To encourage the development of potential contractors for the future.
❑ To deal with a contractor who is already known to the promoter.

Whatever the reason, the promoter should always make a positive decision, for a clear reason. That clear reason should then govern the promoter's decisions on the number, scope, type and terms of the contracts.

## 11.5 Project objectives

The most important decision that the promoter must make in relation to any project is to define precisely what the objectives of that project are to be. This means deciding the answers to a number of questions. The questions are very simple. The answers may be very complex indeed. Some of them are listed below.

❑ What does the organisation want to have?
❑ What does the organisation need to have?
❑ What can the organisation afford to spend on the project?
❑ How will the money spent on the project be recovered?
❑ What can the organisation provide from within its own internal resources?
❑ What does the organisation wish to provide from within its own resources?

❏ What is the time-scale within which the project must be complete?
❏ What are the management resources that the organisation has available to manage the project?
❏ What type(s) of contract are those resources competent to place and manage?
❏ What type(s) of contract does the organisation usually use?
❏ Can the organisation define accurately what it wants/needs to have, or is producing that definition part of the project?

The answers to these questions will tend to define the project and its objectives, and to define them in a way that allows a choice of contract strategy and contract type to be made.

The basic project objective will usually be to produce a usable asset, of one form or another. Other objectives may also be important, or even dominant in some circumstances, such as:

❏ the management and control of risk;
❏ ensuring high standards of quality assurance or plant/process validation;
❏ ensuring high standards of site safety;
❏ ensuring high standards of environmental protection;
❏ assistance by the contractor in operating or maintaining equipment;
❏ the performance of plant;
❏ the long-term operability of plant;
❏ guaranteed supplies of spare parts for plant;
❏ the ability to modify or expand the plant;
❏ training of the promoter's staff;
❏ the acquisition of know-how;
❏ influencing or controlling design by the contractor;
❏ influencing or controlling the identity of any sub-contractors;
❏ retention of construction plant by the promoter;
❏ promoter involvement in project and construction management;
❏ cooperation in construction and implementation;
❏ substitution of labour-based construction;
❏ establishment of an operating/maintenance force.

Project management should bear all the objectives in mind, rank them in importance, and seek a strategy to optimise their achievement.

It is always essential that the project manager ensures that the promoter clearly defines the project objectives. Furthermore, once those objectives have been defined, it is equally important that they are

properly understood by management within the promoting organisation. The likelihood of a successful project is greatly improved when all
the managers of the design, construction, user and support groups within
the organisation are fully informed of the objectives of the project, and
are committed to those objectives.

## 11.6 Contract selection

Figure 11.3 seeks to encapsulate the considerations that must underlie
the choice of contract type that the promoter has to make. Note that at
one end of the scale there are the 'price-based' contracts, while at the
other 'quantities/rates-based' contracts and reimbursable contracts
operate in a broadly similar fashion. A reimbursable contract may
operate on the basis of 'cost + profit' or 'agreed rate' payments. The
principal considerations are explained below.

### *Discipline*

The price-based contract imposes 'discipline' upon the contractor. It
requires a specification to be met, and a price and (usually) a fixed time-
scale within which to comply. Any failure to comply is automatically
penalised by the terms of the contract. Therefore, the contract focuses
the mind of the contractor upon the need to be efficient in carrying out
the work, and the need to provide a result that meets the specification
included in the contract. Anything less and the contractor risks not
making a profit. Therefore, in theory at least, the task of the project
manager is straightforward: simply to observe the progress made by the
contractor and administer the terms of the contract.

The quantities/rates-based and reimbursable contracts impose far less
discipline on the contractor. Even if there is a specification to meet, and
perhaps also a time-scale within which to do so, there are no price
constraints. Indeed, the more chargeable units of work or man-hours
spent in carrying out the contract, the higher the profit made. Therefore,
if 'discipline' is to be imposed on the contractor, it will not come from the
basic contract structure. It will need to be written into the contract or
imposed by the promoter or project manager in such forms as a target–
cost arrangement, detailed contract reporting procedures, or arrangements to control and monitor the contractor's work. (The admeasurement procedure, which is common within civil engineering contracts, is a
typical example of this.)

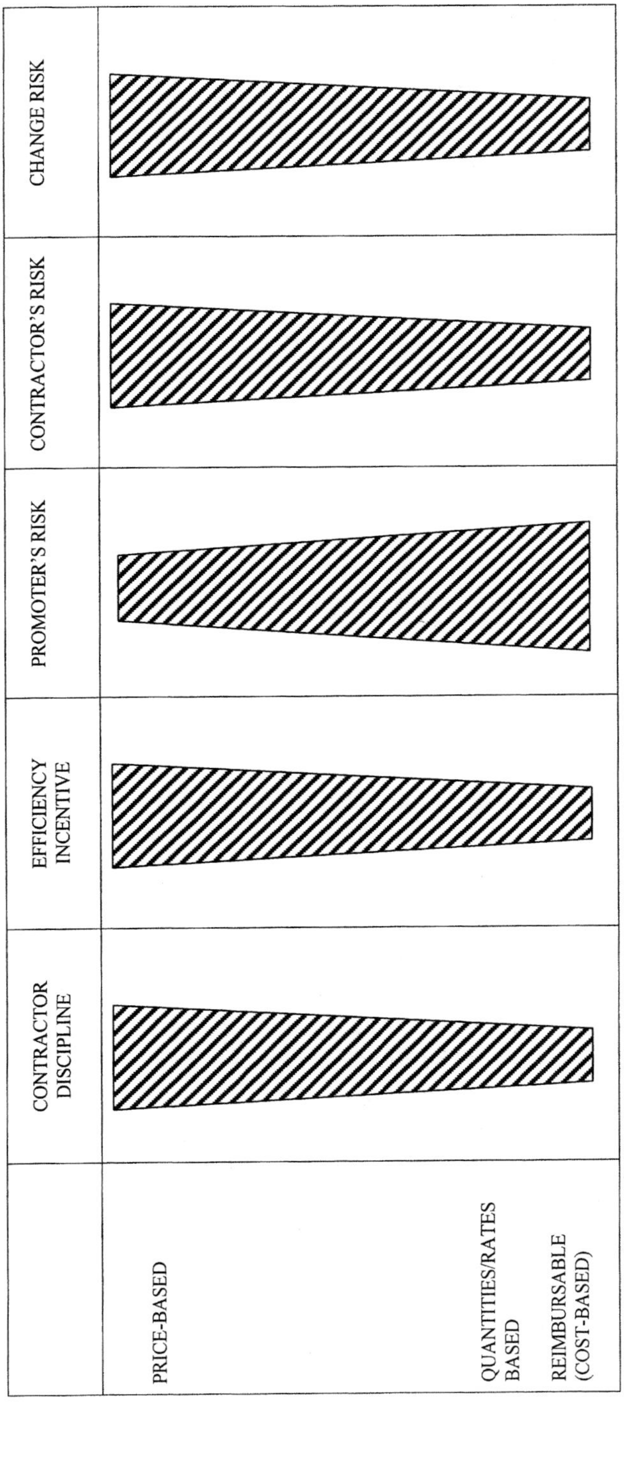

**Figure 11.3** Contract types. The promoter must choose.

## *Incentive*

The price-based contract gives the contractor two incentives. The first, already referred to, is to plan and run the contract efficiently, because this reduces cost and increases profit. Planning, however, creates a potential problem for the promoter in that a carefully planned and organised contract will never result in the shortest delivery period. (Minimising costs means avoiding the problems of overlapping activities, for instance.) A tightly planned contract will also always be vulnerable to severe disruption if work is affected by such problems as force majeure or change.

The second is only to supply the minimum amount of work or equipment necessary to comply with the contract, to 'design, or supply, down to the specification', because this also saves cost. In one sense the inherent risk of the price-based contract is under-design/-supply (just as the inherent risk of the quantities/rates-based and reimbursable contracts is over-design/-supply.) Therefore, the fixed/firm price contract must always be based upon an adequate specification. An adequate specification is one that describes the result required from the contractor sufficiently accurately to minimise any opportunity for significant under-design/-supply.

The quantities/rates-based and reimbursable contracts on the other hand, give the contractor no incentive to under-design/-supply: it gives an incentive to over-design/-supply, since the more work done, the greater the profit. Therefore, what a quantities/rates-based or reimbursable contract should do is to give the promoter proper project management powers to ensure that the contractor performs efficiently.

## *Risk*

The price-based contract automatically imposes a considerable degree of price/money risk on the contractor, and will usually impose a high level of other risk as well. Conversely the promoter carries a low level of financial risk. The promoter knows exactly what the contractor will charge and when payment is needed – and it is normal not to have to pay the full price until after the contractor has met the contract requirements.

Under quantities/rates-based and reimbursable contracts, on the other hand, the contractor usually bears a rather lower degree of price/money risk, and a much higher level of this risk is borne by the promoter.

## *Change*

The great weakness of the price-based contract, from the promoter's point of view, is that as it sets up a rigid contract structure, it may not cope very well with anything more than the minimum degree of change during the life of the contract. This makes the tasks of the project manager and contractor much more demanding if a substantial degree of change does happen. The quantities/rates-based and reimbursable contracts, by their very nature, cope much better, and are therefore inherently more suitable where significant change can be foreseen. However, that change will still need to be managed.

The promoter must also remember the implications of these different types of contract. Some of the more obvious implications that can be identified are listed below.

### Time-scale

By its very nature, the price-based contract requires a longer time-scale to put into place and to carry out than the other types. To place a price-based contract requires the promoter to put together a detailed inquiry document, including in particular a detailed and comprehensive specification. This takes time to prepare. Then the contractor will need time to examine that specification, finalise designs, bring in prices from potential subcontractors, etc., and then put together the tender in response. Finally, the negotiations of the final contract document will almost certainly take more time. As has been shown already, once that contract is in place it will probably require more than the absolute minimum length of time to carry out. The net result is that price-based contracts do take a significant time at all stages. If time (or quality) is the main consideration, the price-based contract is not always the best choice. If cost is the main consideration, however, the time-based contract will always be the best choice provided that the contract can incorporate an adequate specification *and* that the level of voluntary change is low.

### Relationship

In the fixed/firm price contract, the relationship must always be at 'arm's-length' even when it is highly cooperative. In the reimbursable contract, the relationship should be much more collaborative. This is a much more sophisticated relationship and requires a very different style of project management on both sides. In particular, it requires more, and higher quality, project management time from the promoter.

### Change

It has already been shown that the price-based contract copes badly with change. From the point of view of the contractor, implementing change part-way through the contract within the context of a tightly planned contract has considerable time, cost and organisational implications. This means that an arm's-length relationship between the two parties will always come under a certain amount of stress if the managers on both sides find themselves in a series of negotiations about the cost and time implications of changes imposed on the contract by the promoter. It is all too easy for the arm's-length relationship to become both adversarial and abrasive unless it is carefully managed.

Change is much easier to implement within the context of a reimbursable contract, since the payment structure of the contract means that the cost/profit-recovery problem for the contractor is minimised. Change is therefore much less adversarial. Indeed the problem is that change can become too easy, and therefore too tempting, for the promoter, so that too many changes are made and the cost of the project becomes too high.

In recent years, a particular problem has arisen with changes in quantities/rates-based contracts within the UK construction industry. The UK civil engineering and building industries have experienced the growth of a 'claims culture'. Many contracts have become the subject of adversarial claims for extra payments, which consume considerable resources on both sides before they are settled. This claims culture is directly contrary to theory. (Change within the quantities/rates-based contract should be no different to change within the reimbursable contract.) It was to some extent instrumental in the production of the ECC conditions (see below).

## 11.7  Project organisation

The number and sequence of contracts for equipment and services can vary greatly from project to project, depending upon the capabilities or preferences of the promoter (Figure 11.4). Often a consultant or design contractor may be employed to advise upon the feasibility stage of a proposed project, or a project management consultant to advise on risks and contract strategy. For implementation, it may be appropriate for a promoter to employ a single contractor for the whole project. Or a project might be so large or so diverse that no one contractor is appropriate to share the risks. Then that project might be let to a consortium of companies, or be shared out among two or more separate contractors. For the equipment required for a new factory, one contractor might be employed to install equipment supplied by others.

A series of contractors can be employed in turn for construction work,

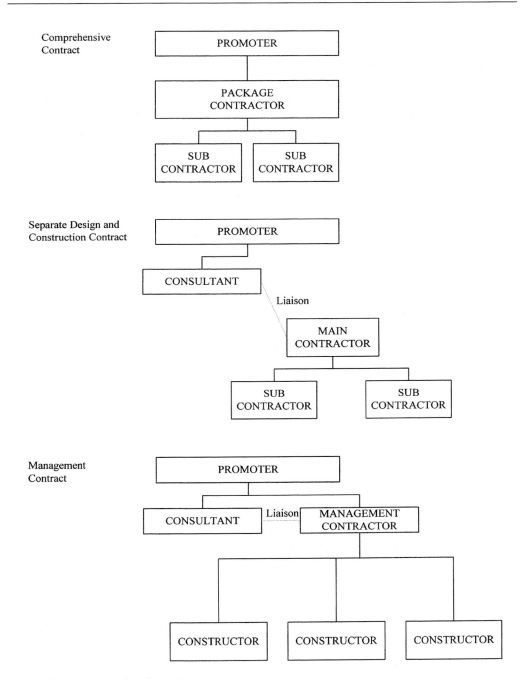

**Figure 11.4** Organisational structure.

for instance one for demolition work, another for new foundations, the superstructure and building work, and others for designing and supplying equipment, installing it, completing systems and services, or testing and commissioning for replacing part of a factory, each under different terms of contract. For a building project, different contractors could be employed separately for the structural, finishing and services work, instead of one main contractor who would sub-let these specialist tasks. Alternatively a joint venture or consortium approach could be adopted.

The choice of an appropriate organisation for any project should consider:

❑ the ability to meet project objectives;
❑ the resources and services offered;
❑ the resources and expertise the promoter is able to commit to the project;
❑ the balance between management, design and implementation.

The common types of project organisation are described below.

### Package deal ('turnkey', design and supply)

The simplest arrangement contractually is that a single main contractor is responsible for carrying out everything necessary for the project from the start to the hand-over of the completed work to the promoter. Payment is generally on a lump sum or reimbursable basis, although this is often broken down into elements or phases of work. Although the main contractor is responsible for engineering, procurement, construction, installation of equipment, and testing and commissioning, parts of the work may be sub-contracted to specialists. The main contractor may also be responsible for ancillary activities such as financing, obtaining public approvals, purchasing process materials or other functions.

The main strengths of the package deal contract are:

❑ pricing/costing the project at an early stage may be possible provided the promoter's requirements are known;
❑ the price/cost of the project will probably be reduced;
❑ the time can be shortened as a result of design/construction overlap;
❑ design integration is much easier;
❑ overall project organisation tends to be better;
❑ the promoter has to deal with one organisation only for both design and construction;

- fewer promoter's resources need to be involved;
- the design should be tailored to give the most efficient construction;
- operating requirements are better dealt with;
- the causes of defects are less a matter of dispute.

However, there are some weaknesses, which primarily relate to the role of the promoter in the project. These include:

- the contractor may not always do what the promoter expected, and the promoter's ability to control the contractor will be fairly low;
- managing the contractor will require high-quality skills;
- the promoter must commit to the whole package at an early stage;
- the promoter will have little contact with sub-contractors;
- the promoter is in a relatively weak position to negotiate change;
- the extent of competition is likely to be reduced;
- some in-house practices of the promoter might constrain the contractor.

In the process/plant and service industries, where contractors are widely experienced in this type of contract, package deal contracts are widely used. In such fields as civil engineering, the package deal contract may be used when the building is of a standard or repetitive nature, for example houses, warehouses or office blocks, and it can also be used where contractors offer specialist design/construction expertise for particular types of project. It can also satisfy a need for an early start to a project. When the promoter and any advisors have insufficient specialist management resources for the project, or if the promoter wishes to place the work with a single organisation, this form of contract is very effective.

### Concession contracts or build–own–operate–transfer

The build–own–operate–transfer (BOOT) concept can be defined as a major start-up business. The contractor, or consortium of contractors, undertakes to build, own and operate a facility that would normally be undertaken by a government or a private promoter organisation for a fixed concession period. The ownership of the facility then returns to the government or private promoter organisation at the end of that concession period. These contracts are described in more detail in Chapter 17.

## Separation of design and implementation

The separation of design and implementation results in a series of two or more contracts for successive stages of the work for a project. A common example is the employment of an architect or consulting engineer for the design of a project, and then a main contractor for its construction/ implementation. This is still the conventional or traditional contract arrangement for many building and civil engineering projects within the UK. Often the design contractor will supervise the construction/ implementation contract on the promoter's behalf. Construction/ implementation is usually undertaken under a quantities/rates-based contract, or occasionally under a lump-sum or reimbursable contract. Management responsibilities are divided.

## Management contracting

Management contracting is an arrangement whereby the promoter appoints an external organisation to manage and coordinate the design and construction phases of a project. The management organisation itself does not normally execute any of the permanent works. These are packaged into a number of discrete contracts, but may provide specified common user and service facilities.

When management contracting is used, the promoter is creating a contractual and organisational system which is significantly different from the conventional approach. The management contractor becomes a member of the promoter's team, and the promoter's involvement in the project tends to increase.

Payment for the management contractor's staff is reimbursable plus a fee, but the engineering contracts, which are arranged by the management contractor, will usually be lump-sum or quantities/rates-based. The management contractor is appointed early in the project's life and has considerable design input. Designers and contractors are employed in the normal way. Another benefit is that the work can be divided into two or more parallel simultaneous contracts, offering reduced duration and risk sharing. Separate contracts for independent parts of the construction of a project divide a large amount of work amongst several contractors.

Some contracts will be consecutive, but it is possible to award two or more parallel simultaneous contracts. Examples in industrial construction and maintenance are separate contracts to employ the experience of different contractors' on different types of work.

Management contracts have considerable advantages in large civil engineering and building contracts, where the package deal contract cannot normally be used. The main advantages are listed below.

- Time saving, resulting from the extensive overlapping of design and engineering, is one of the main advantages of this form of contract. By saving time, the impact of cost inflation can also be reduced.
- Cost savings can also arise from better control of design changes, improved buildability, improved planning of design and engineering into packages for phased tendering, and keener prices owing to greater competition and better packaging of the work to suit the contractor's capabilities.
- Further time saving can arise from the special experience of the management contractor, for example overcoming shortages of critical materials, and delays arising from a scarcity of particular trades.
- It provides flexibility for the promoter's design changes during construction to improve the fitness of the project for his needs, and the contractor's design changes to improve buildability.
- Discipline can be imposed on the promoter's decision-making procedures and on the management of design, in particular to ensure proper coordination of the availability of design information with the sequence of work, especially on multi-disciplinary projects.

There are also some potential drawbacks with management contracting.

- The promoter may be exposed to greater risk from the contractors than in a conventional arrangement.
- There is some risk in the absence of an overall tender price for the complete works at the start of the project.
- There is a tendency to produce additional administration and some duplication of supervision staff.

Management contracts can be particularly advantageous when there is a need for:

- an early start to the project for political reasons, or budgetary or procurement policy;
- an early completion of the project, but the design cannot be completed prior to construction (this situation requires good planning and control of the design/construction overlap and careful packaging of construction contracts, which is the normal skill of the management contractor);

❑ innovative and high-technology projects, when it is likely that design changes will occur throughout;
❑ organisational complexity, which typically may arise from the need to manage and coordinate a considerable number of contractors and contractual interfaces, and possibly several design organisations, and it is useful when the promoter and advisors have insufficient specialist management resources for the project.

### Offshore oil engineering

The offshore oil industry employs the same range of contract types as those used in other industries, with the exception of the Concession contract, which is not used. However, the industry has always used totally different terminology to other industries when classifying its contracts. Contracts within the industry are usually defined in terms of acronyms based upon the work to be carried out by the contractor. Thus we have the engineer (i.e. design), procure, construct contract (EPC), the procure and construct (to a design supplied by the promoter) contract (PC), the engineer, procure, install, commission contract (EPIC) and the procure, install, commission contract (PIC). In broad terms, the EPIC contract is equivalent to what other industries would describe as a turnkey contract.

The only substantial difference between offshore oil contracts and others is that oil industry contracts, even EPIC contracts, always provide for a high level of promoter involvement at all stages of the work.

Project services and management services contracts have been developed to deal with very large and complex projects such as oilfield development. These operate in a similar way to management contracts in building and civil engineering, but with some significant organisational differences, for example:

❑ very large numbers of staff are involved, and 'management of management' becomes crucial to the success of the project;
❑ the project services contractor is formally integrated into the promoter management structure.

### Direct labour

Instead of employing an external organisation, a promoter may have equipment made or installed, or some projects constructed by an in-house maintenance or construction department, which is known in the UK as 'direct labour' or 'direct works'. If this is the case, in all but the smallest organisations the design decisions and the consequent manufacturing, installation and construction work are usually the responsibility of different departments. To make their separate responsibilities clear, the order instructing work to be done to the design may, in effect, be the equivalent of a contract that specifies the scope, standards and price of the work as if the departments were separate companies.

This arrangement should have the advantage of clarifying the responsibilities for project costs due to design and those due to the consequent work. These internal 'contracts' may be very similar in outward form to commercial agreements between organisations, but of course any dispute between the two departments would be settled by a managerial decision within the organisation rather than by external legal or other dispute resolution procedures. Contractual principles and these notes should therefore be applied to them.

As described in chapter 17, if the contractor both promotes and carries out a project, then consideration should be given to separating the two roles. The expertise and responsibilities involved in deciding whether to proceed with a project are totally different from those required when implementing that project. Separation of these responsibilities may also be required when other organisations are involved, for example because they are participating in financing the project. For all such projects, except small ones, an internal contract may again be appropriate to define responsibilities and liabilities.

## 11.8  Risk allocation

A prime function of the contract is to allocate, between the parties, both the responsibilities and liability for risk. Risk may be defined as possible adverse consequences of uncertainty. The identification of potential areas of risk/uncertainty and consideration of the appropriate way to manage them is a logical part of the development of organisational and contractual policies for any project. Some of these uncertainties will remain whatever type of contract is adopted, and the tender must include a contingency sum for them.

Some risks can be due to a promoter's inability or unwillingness to specify what is wanted at the start of the project. If, for instance, the contract is to be for the design and supply of equipment that is to be part of a system, and that system cannot properly be defined until after work on the contract has begun, or if the initial prediction of the purpose of a project has to be varied during the work in order to meet changes in forecasts of the demand for the goods or services that are to be produced, then a high level of uncertainty will be implicit in the project.

Promoters rarely invite tenders on the basis that an organisation is to be committed to complete the work regardless of risks. Contractors would then have to cover themselves by prices far in excess of the most probable direct and indirect costs they might incur. Governments and other promoters in countries with less engineering expertise do, at times, ask bidders to carry very high levels of risk, but the trend in industrialised countries since the early twentieth century has always been that a risk should be the responsibility of whichever party is best able to manage it to suit the objectives of the project.

All parties to a project carry risk to some extent whatever the contracts between them, because, for example, that work may be frustrated by forces beyond their control. If so, the time lost and all or some of their consequent costs may not be recoverable. The basic division of risks between the parties is usually established in the conditions of contract specified by the promoters when inviting bids.

## 11.9  Terms of payment

| | |
|---|---|
| *Price-based:* | 'lump-sum', 'fixed' or 'firm' prices are stated in the contract |
| *Quantities/rates-based:* | the contract contains a bill of quantities (BOQ) or schedule of rates. |
| *Cost-based:* | cost-reimbursable, either on the basis of actual cost (plus a profit) or agreed unit rates (sometimes combined with a target cost mechanism). The actual costs incurred by the contractor are reimbursed, together with a fee for overheads and profit. |

### Lump sum

A lump-sum payment is based on a single tendered price or fee for the whole of the work. Again it is important to remember that different industries and organisations use the terminology in totally different ways. Like many other words used in contract management, words can be used to mean different things. What matters in each contract are the terms of payment in that contract.

The words 'lump sum' are used in civil engineering to mean that the contractor is paid on completing a major stage of work or reaching a milestone, for instance on handing over a section of a project. In the process industry, on the other hand, the term 'lump sum' is used to mean that the amount to be paid is fixed, but perhaps subject to change as a result of escalation.

The words 'firm' and 'fixed' can also be given different meanings. Some industries use the word fixed to mean that the contract price is to be the final price because it will not be subject to escalation, while a firm price will be subject to escalation. Others use the words in the opposite way.

Payment to a contractor under a price-based contract will usually be in stages; in a series of lump sums each paid upon his achieving a milestone, i.e. a defined stage of progress. The use of the word milestone usually means that payment is based upon progress in completing what the promoter wants. Payment based upon achieving defined percentages of a contractor's programme of activities is also known as a 'planned payment' scheme.

Although relatively inflexible, contracts will generally include a 'variation clause', which provides the promoter with the right to make changes within the scope of the works, coupled with a negotiation of consequent price/time/specification changes.

The advantages to the promoter of using a price-based contract are:

❏ a well-understood and widely used type of contract in the equipment plant process and offshore industries;
❏ a high degree of certainty about the final price;
❏ easy contract administration provided there is little or no change;
❏ it facilitates competitive bidding by contractors;
❏ the promoter's management resources are freed for other projects.

The disadvantages are:

❏ it is unsuitable when a high level of change is expected;
❏ there is always the possibility that a contractor who has accepted too low a price may be faced with a loss-making situation, especially where considerable risk has been placed with them, and this may lead to cost cutting, claims and, in extreme cases, the collapse of the project;
❏ the promoter and advisers will have minimal involvement/influence upon the implementation of the project by the contractor.

Lump-sum contracts provide an incentive for the contractor to perform welll when design is complete at tender and little or no change or risk is envisaged. It can be adopted if the promoter wishes to minimise the resources involved in contract administration. It should be noted that this type of contract is rarely used for main civil engineering contracts. However, it is much more common in process plant contracts, and is virtually standard practice within the equipment industries.

### Quantities/unit rate (admeasurement)

Admeasurement is based on 'bills of quantities' (BOQs) or 'schedules of rates', and is typically used in civil engineering and building contracts. Items of work are specified with quantities. Contractors then tender rates against each item. Payment is usually monthly, and is derived from measuring quantities of completed work and valuing those quantities at whatever rates are finally agreed in the contract.

Quantities/unit-rate contracts provide a basis for paying the contractor in proportion to the amount of work completed, and also in proportion to the final amounts of the different types of work actually required by the promoter if this is different to the amount predicted at the time of agreeing the contract.

Within the UK, the predicted amounts are usually stated in a bill of quantities which lists each item of work to be done for the promoter under the contract, for example the quantity of concrete required, to a quality defined in an accompanying specification. An equivalent in some industrial contracts is a 'schedule of measured work'. In these contracts, the total amount to be paid to the contractor is based on fixed rates, but changes as the quantities change.

In some contracts, what is called the 'schedule of rates' is very similar to a bill of quantities in form and purpose. When bidding, contractors are asked to state rates per unit for all items on the basis of an estimate of possible total quantities in a defined period, or within a limit of, say, $\pm 15\%$ variation on these quantities. In other cases, the rates are to be the basis of payment for any quantity of an item which is ordered at any time.

The principal strengths of the quantities/rates-based contract are:

- ❑  a well-understood, widely used type of contract within the civil engineering and building industry;
- ❑  some flexibility for design change;
- ❑  some overlap of design with construction;
- ❑  good competition at tender;
- ❑  the tender total gives a good indication of the final price where the likelihood of change, disruption and risk is low.

The principal weaknesses of the quantities/rates-based contract are:

❑ difficult claims resolution, which is quantity-based and adversarial;
❑ limits to flexibility, since new items of work are difficult to price;
❑ limits to the involvement of the promoter in management;
❑ the final price may not be determined until long after the works are complete, especially when considerable change and disruption has occurred and major risks have materialised.

An admeasurement contract can be used with a separate organisational structure. It requires that the design is complete, but can accommodate changes in quantity. It is used on many public sector civil engineering projects such as roads and bridges where little or no change to the programme is expected, and the level of risk is low and quantifiable.

It should be noted that admeasurement contracts are sometimes used on high-risk contracts or where considerable change and disruption are expected, but the promoter's procedures and regulations prevent the use of a cost-based contract. In these circumstances, promoters should proceed with caution and note the deficiencies of the traditional bill of quantities approach in allowing for extensive change, and particularly for disruption to the programme.

### Cost reimbursable

Cost-reimbursable contracts are based on the payment of the actual cost plus a specified fee for overheads and profit. The contractor's cost accounts are open to the promoter (open-book accounting). Payments may be monthly in advance, in arrears, or from an imprest account.

The simplest form is that of a contract under which the promoter pays ('reimburses') all the contractor's actual costs for employees employed on the contract ('payroll plus payroll burden'), and of materials, equipment and payments to sub-contractors, plus usually a fixed sum or percentage for financing, overheads and profit. Some items, such as computer time or telecommunication costs, will be paid for at unit rates. Agreed low-profit rates will often be used as the basis for reimbursement of the contractor's employee costs as well. Sometimes the costs of satisfactory or acceptable work only are reimbursed, and none, or only some, of the costs of rejected work.

Contractors working under this low-risk arrangement should not expect to be rewarded at the same rate of profit as under price-based contracts because the risk burden is less.

In project management terms, the cost-reimbursable contract probably tends to produce the most sophisticated and collaborative relationship between promoter and contractor of these three types of contract. Within the processing and offshore industries it is widely used in large projects, where the detailed design/specification of the plant does not exist at the date of contract. It is also used for 'fast-track' contracts, and as the basis for 'partnering' contracts. Finally, it is also used in contracts for development, repair and maintenance work, and a range of other work where the scope, timing or conditions of work are uncertain. Cost-reimbursement is therefore the basis of payment in many 'term contracts' to provide construction or other work when ordered by a promoter at any time within an agreed period (the 'term').

Under all such contracts, the promoter in effect employs a contractor as an extension of their own organisation. To control the cost of cost-reimbursable work, the promoter's project team must direct the contractor's use of resources. The contractor's risks are limited, but so is the prospective profit.

The advantages of cost-reimbursable contracts are:

❑ extreme flexibility, plus a high degree of collaboration between parties;
❑ fair payment for, and good control of, risk;
❑ they allow and require a high level of promoter involvement;
❑ they facilitate joint planning;
❑ a knowledge of actual costs is of benefit to the promoter in estimating and control, and in evaluating proposed changes.

The disadvantages of cost-reimbursable contracts are:

❑ little incentive for the contractor to perform efficiently (but see below);
❑ no estimate of final price at tender;
❑ administrative procedures may be unfamiliar to all parties; in particular, the promoter must provide cost accountants or cost engineers who must understand the nature of the contractor's business.

In one sense, cost-reimbursable contracts are weak contracts from the promoter's viewpoint. Unless both promoter and contractor are highly professional in their approach, they should be restricted to contracts containing major unquantifiable risk, and to projects where no other form of contract is feasible.

There are a number of situations where the use of this type of contract can be justified, i.e. when the work is of an emergency nature, or when the work is innovative and productivities are unknown, for example work involving research and development. Cost-reimbursable contracts can be used if the work is of exceptional organisational complexity, for example when there are multi-contract interfaces, to the extent that a definition of the target cost is impossible.

*Target cost*

Target cost contracts are a development of the reimbursable type of contract. The promoter and contractor enter into a reimbursable contract, but also agree a probable (or target) cost for the probable scope of work, together with an incentive payment mechanism which deals with any difference between actual outturn and target cost. The contractor's actual costs are monitored and reimbursed under the reimbursable contract. Any difference between the final actual cost and the final target cost is then shared by the promoter and the contractor in accordance with the incentive mechanism. Before reaching the stage of tender documents, the promoter is recommended to investigate the implications of several different incentive mechanisms. If the incentive is to be maintained, the target cost will subsequently be adjusted for major changes in the work and cost inflation. A fee, which is paid separately, covers the contractor's overheads, and any other costs specified in the contract documents as not being allowable under actual cost and contractor's profit.

The following simple example and Figure 11.5 show the effects of a 50/50 share incentive mechanism and a fixed fee on the payment by the promoter and on the financial incentive to the contract. The target cost is £150 million, the difference between the actual cost and the target cost is shared 50/50 between the promoter and the contractor, and the fixed fee is £15 million.

In a case where the actual cost equals £135 million, the total payment by the promoter is £157.5 million, which is less than the sum of £165 million which he would have paid if the actual cost had equalled the target cost. The contractor receives a lower total payment, but his margin, as a percentage of actual cost, is increased from 10% to 16.7%. Conversely, if the actual cost exceeds the target cost, the promoter is partly cushioned by the incentive mechanism. The cushioning effect also applies to the contractor, but since he has to bear 50% of the excess costs, his margin is much less attractive (4.5%). This illustration clearly demonstrates two powerful features of a target cost contract: first, the motivational effect of the incentive mechanism on the contractor and second, the benefits to both parties of working together to keep the

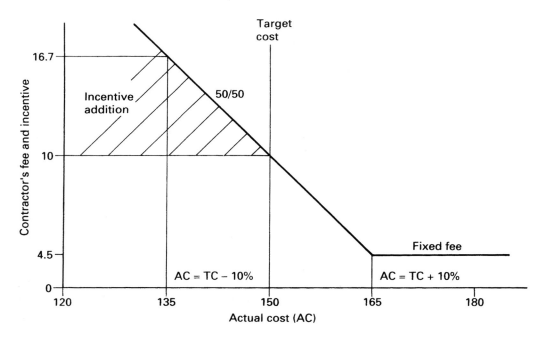

**Figure 11.5** Contractor's fee relative to actual cost.

actual cost under strict control and to bring it below the target cost whenever possible. Thus, the target cost mechanism remedies the principal weakness of a pure cost-reimbursable contract by imposing an incentive on the contractor to work efficiently.

The advantages of this form of contract are:

- fair payment for, and good control of, risk;
- a high level of flexibility for design changes;
- identity of interest, i.e. both parties have a common interest in minimising the actual cost, fewer claims result and settlement is easier;
- promoter involvement, i.e. the contract offers an active management role for the promoter or their agent, and joint planning aids the integration of design and construction, the efficient use of resources and the satisfactory achievement of objectives;
- realism, i.e. the facility to require full supporting information, with bids coupled with thorough assessment, to ensure that resources are adequate and the methods of construction are agreed;

❑ knowledge of actual costs, which is of benefit for estimates and control, and in evaluating proposed changes.

There are some potential problems in using this type of contract which might make it unsuitable for certain projects. Some important points to consider include:

❑ promoter involvement is essential, and a different attitude must be taken from that adopted on price-based contracts;
❑ unfamiliar administrative procedures and a probable small increase in administration costs.

The use of this form of contract should be considered when there is an inadequate definition of the work at the time of tender owing to the emphasis on early completion or an expectation of a substantial variation in work content, when the work is technically or organisationally complex, or when the work involves major unquantifiable risks. The role of the promoter is significant and a target contract can be used if the promoter wishes to be involved in the management of the project, or wishes to use the contract for training of his own staff or for the development of a local skilled construction labour force.

Targets can be based on other project parameters instead of, or as well as, cost. Time-target contracts are popular for certain types of work and Figure 11.6 shows a graphical representation of a combined time- and cost-target mechanism.

## 11.10 Model or standard conditions of contract

The great majority of projects will use contracts that are governed by 'standard' or 'model' sets of contract conditions. By standard contract conditions, we mean 'in-house' conditions. In-house conditions have the advantage, or disadvantage, that they will always be biased to a greater or lesser extent in favour of the company that writes them. By 'model' conditions, we mean industry-wide conditions, which are generally accepted as a fair or reasonable basis for contracts within a particular industry.

The advantages of using standard and model conditions are obvious. It saves enormously on time and resources if the contract can be based on pre-existing contract conditions that are already known and understood by both sides. If both sides understand the conditions from the start, it is easier and quicker for the contractor to tender, and easier for both sides

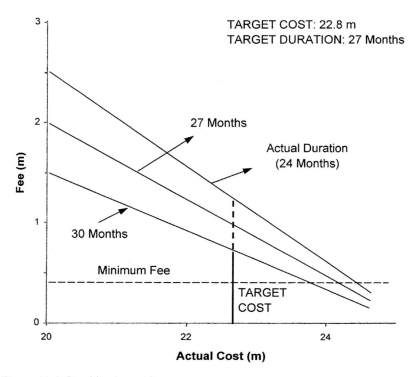

**Figure 11.6** Combined cost/time target.

to manage the contract. Model conditions have the added advantages that contract negotiation will be easier and less adversarial, and that fair conditions generally produce a better contract relationship and probably a more equitable result. In addition, model conditions generally have accepted and understood meanings. They have been used before in hundreds or thousands of contracts. They have been debated by lawyers and managers in dispute after dispute. Their meaning is well established.

Promoters and contractors need to understand how such conditions come to be written and how they operate. No set of standard or model conditions exists in isolation. Every set is written by a committee. Standard conditions will generally be written by a committee drawn from within the organisation. Model conditions are written by committees that represent the various interests within the industry. However the committee is made up, it is given the task of writing a set of conditions that is suitable for a *typical* contract. In the case of standard conditions, it may be for a typical contract within a particular company. In the case of model conditions, it will be for a typical contract within the industry as a whole.

The conditions will then be written to describe the work that the parties will usually carry out in that kind of contract, and the various necessary stages of that type of contract, such as design, manufacture of equipment, site preparation, construction, etc. The conditions will also allocate the various commercial, technical and economic risks involved between the parties to the contract, and provide appropriate solutions to the more predictable problems that may arise.

The consequence of this is that the different sets of conditions, especially model conditions, often deal with completely different issues and problems. Civil engineering contracts, for instance, deal with contracts where almost all the work is carried out on the site. They deal with the problems of managing a large complex site operation and of dealing with such issues as 'bad ground' conditions. On the other hand, contracts for the supply of electrical or mechanical equipment, such as a computer system, deal with completely different problems. The contractor will spend many months designing the system, manufacturing equipment, writing and testing software, etc. Then everything is taken to the site, and half-a-dozen people can install the equipment in a few days. A building contract might cover the erection of a building; the building must be constructed to a good standard, and it will be inspected to ensure that this is so. A process industry contract might cover the design and supply of a plant to manufacture pharmaceutical or food products. The plant will be inspected to ensure that it is properly constructed, but the design/ construction would then also need to be validated, and the plant will be subjected to a series of exhaustive operating and performance tests to prove that it can manufacture the correct products for the correct cost.

Different types of commercial/contract situation require different conditions of contract because the work is different, the problems are different, and even more important, the risks are different. To understand any set of model conditions, and therefore to be able to manipulate and make the best use of them in practice, the promoter needs to understand the points listed below:

❏ what the basic commercial/contract situation is which that set is principally designed to cover. In other words, we have to read through the conditions and ask ourselves 'What kind of story are the conditions telling us?'
❏ how flexible the conditions are, and therefore how easily and how far they can be adapted to cover other commercial/contract situations.
❏ the commercial attitude (or, almost, the bias) of the conditions. Do they favour the promoter or the contractor, and if so how? Do they try to hold some sort of balance between the two? Do they treat the

parties as having an equal degree of skill, or do they treat one party as having a lower degree of skill than the other?

❑ what the more important sections of the conditions actually say; in other words, how the conditions actually operate in practice.

Promoters can, if they wish, use in-house contracts or commission new contracts for a specific project. Sometimes this will be the only appropriate course to take, simply because the project is so unusual that no existing conditions are appropriate. However, there are drawbacks with this approach: primarily that the contract will not have been 'tested at law', and the precise meaning and interpretation of any clause in the new contract could potentially be the source of lengthy litigation. It is more likely that the promoter will select a set of conditions of contract that most closely matches the requirements of the project, and then modify some clauses or add additional 'special' clauses as necessary. Even then, however, the meaning of the modified or additional clauses may be open to question.

From time to time there have been attempts to produce model conditions of contract to bridge the gaps between different industries. This would allow consistent risk and incentive strategies to be applied across multi-disciplinary boundaries. The most recent attempt in the UK has been the Engineering and Construction Contract (ECC), originally published as the New Engineering Contract in 1993. ECC was sponsored by the Institution of Civil Engineers, and was intended for all types of engineering and construction, under any contract strategy, in any country and under any legal system.

## 11.11 Sub-contracts

The same contract strategy principles apply to decisions made by a main contractor to employ sub-contractors. Hence the section above can be interpreted at a number of levels, with different parties filling the 'promoter' and 'contractor' roles. For example, in a sub-contract the contractor fills the promoter role and the sub-contractor fills the contractor role, but in a further supply contract the sub-contractor could fill the promoter role and the vendor fill the contractor role.

A common principle is that in a main contract a contractor is responsible to the promoter for the performance of his sub-contractors. Practice varies as to whether a main contractor is free to decide the terms of sub-contracts, choose the sub-contractors, accept their work and decide when to pay them. It also varies as to whether or when a promoter may by-pass a main contractor and take over a sub-contract. In nearly all engineering and construction, the main contractors employ sub-contractors and suppliers of equipment, materials and services in parallel.

## Reference

Barnes, N.M.L. and Wearne, S.H. (1993) *The Future for Major Project Management*, Proceedings of the 1st British Project Management Colloquium, Henley.

## Further reading

Bailey, P. (1987) *Purchasing and Supply Management*, 5th edn, Chapman Hall, London.

Institution of Civil Engineers (1995) *The New Engineering Contract*, 2nd edn, Thomas Telford, London.

Institution of Civil Engineers (1999) *Conditions of Contract*, 7th edn, Thomas Telford, London.

Latham, Sir M. (1994) *Constructing the Team: Final Report of the Government/ Industry Review of Procurement and Contractual Arrangements in the UK Construction Industry*, HMSO, London.

Turner, J.R. and Simister, S.J., eds (2000) Roles and responsibilities. In: *Gower Handbook of Project Management*, Gower, Aldershot, Chap. 30, pp. 571–88.

# Chapter 12

# Contract Policy and Documents

Any organisation that wishes to employ others to undertake work on its behalf must have a coherent policy for their selection, together with appropriate procedures for placing and managing contracts with the contractors that it selects. Possible strategies for the identification and selection of the most appropriate contractor(s) and the broad principles that apply to the selection process are considered and reviewed in this chapter.

## 12.1 Tendering procedures

Several different procedures may be used for selecting vendors, suppliers, tenderers and contractors.

(1)  *Competitive:* open or select (a restricted number of bidders).
(2)  *Two-stage:* a bidder is selected competitively early in the design process. The tender documents contain an outline specification or design, and approximate quantities of the major value items. As design and planning proceeds, the final tender is developed from cost and price data supplied with the initial tender.
(3)  *Negotiated:* usually with a single organisation, but there may be up to three.
(4)  *Continuity:* tendering competitively on the basis that bidders are informed that the successful party may be awarded continuation contracts for similar projects based on the original tender.
(5)  *Serial:* the bidder undertakes to enter into a series of contracts, usually to a minimum total value. A form of standing offer.
(6)  *Term:* the bidder undertakes a known type of work, but without knowing the amount of the work, for a fixed period of time.

In all cases a pre-qualification procedure may be adopted.

## 12.2  Contracting policy

It is important that every promoter involved in contracting has a formalised contracting policy established by management, broadly defining why, what, when and how work should be contracted out by the promoter.

The objective of going out to contract is to obtain specific works, equipment, services and goods required to support the promoter's general business. The promoter will, of course, provide some of these requirements from within their own resources. However, there may be very good commercial reasons why the promoter should obtain them from others, even if the promoter is able to provide them internally. Perhaps a contractor can supply items at lower cost/risk than the promoter. Perhaps a contractor can provide staff, labour and expertise that cannot be made available from within the promoter's own resources, or cannot be made available in the quantity or time required. Perhaps the type, quality, enhancements and fluctuation of the required works, services or goods are inherent to the contractor's special skills and not to the promoter's. Contracting out should encourage the optimum utilisation of the promoter's own executive staff, and keep administrative overheads to a practical minimum.

As a general rule, contractors should be independent, self-sufficient and 'at arms length', and the promoter's aim should be to manage the contract, not the contractor.

Matters which merit review for a consistent policy approach include:

❏   qualification of contractors in qualities such as reputation, experience and reliability;
❏   qualification of contractors in product, location, price/cost, previous performance, etc.;
❏   circumstances under which negotiated, rather than competitive, tenders may be appropriate and/or acceptable;
❏   types of contracts to be preferred in given circumstances;
❏   facilities and services which may be provided to contractors by the promoter;
❏   the promoter's project/contract management resources and skills;
❏   commercial aspects to be included in the contracts themselves.

*Procedures*

It is always important to remember the basic principles. Contract procedures are written for the purpose of outlining the objectives and scope of contract policy and the manner in which it is to be applied. They are designed to ensure that:

❑ the promoter's policies and procedures are clearly stated and applied;
❑ individual authorities and responsibilities for the preparation, award and control of contracts are defined;
❑ appropriate input from advisory and client functions is incorporated at each stage of the contracting process;
❑ information is passed in an orderly manner to all parts of the organisation, as well as to and from appointed contractors, on a need-to-know basis;
❑ the best interests of the promoter and its staff are protected and safeguarded.

*Business ethics*

Management should establish a code of conduct, which should regularly be brought to the attention of all staff, and should cover:

❑ the general policy to be adopted in relationships with contractors and their staff, who should be dealt with in an equitable and business-like manner;
❑ the obligation to declare any conflicts of interest, potential as well as actual, should they arise;
❑ the importance of confidentiality and security in all matters concerned with contracting activities.

Contractors should also be made aware of the promoter's policies with regard to conflicts of interest and the giving and receiving of gifts and so on, and promoters may consider it appropriate to incorporate a suitable clause on the subject in their general conditions of contract.

## 12.3 Contract planning

The promoter will usually differentiate between 'projects', and 'purchasing/procurement'. Every organisation constantly buys goods equipment and services as a normal part of its operations. Sometimes the promoter will simply buy materials for consumption, such as raw materials, fuel or office supplies. Sometimes the promoter will buy spare parts or replacements, or servicing/maintenance work for equipment, plant or facilities. Sometimes contracts will be placed for minor or major modifications to equipment, plant or facilities. Sometimes the purchase of major works or facilities will be required. Obviously the same *principles* must apply to all of these activities. They all involve the use of a contractor to supply some thing or some service to the promoter, and the contractor should be treated on the same general basis, whatever the supply. However the *practices* will change. At some point the promoter will stop purchasing items of equipment or work, and begin to operate on a project basis. That point will change from organisation to organisation, and will often depend upon the personnel involved and upon the circumstances of the case. This book is concerned with the overall principles and the practices that should apply when a project is involved.

During the initial phases of a project, discussions should take place, under the direction of the project manager, between the various departments and disciplines concerned to formulate, review and develop the different aspects of the project, culminating in the preparation of a coherent contracting plan for the project. It will often be necessary to prepare a number of plans to cover the details of each contract to be let. In particular, 'user' departments should *always* be involved in drawing up the project plan.

An overall schedule should be kept to monitor the critical dates in the preparation and letting of contracts over the whole project, and also the critical dates within the sequence of activities for letting each individual contract. As discussions proceed, attention will need to be given to the various implementation considerations, including those factors which will influence the type and number of contracts, selection of possible contractors, and method of operation.

The contracting strategy must address the choice of the number and types of contract that will best contribute to the success of the project, bearing in mind the amount of the promoter's resources that would need to be committed. The types of contract chosen also affect the master plan because of the varying requirement for control and monitoring.

Competitive tendering would be the normal method adopted for letting contracts, and the contract should be awarded to the most acceptable tender provided other criteria of comparability and acceptability are met. Often the most acceptable tender will be the lowest-priced tender. However, that will not necessarily be the case where time-scale, or where, particularly in the process/equipment industries, the quality or performance of equipment may be more important than the price.

### Number of contracts

The first step in contracting planning is to decide the number of contracts into which a project will be divided. One of the basic considerations must be the effect of the number of contracts on the promoter's management effort. More contracts mean more design risk and more interfaces for the promoter to manage, and a greater degree of management involvement. Fewer contracts reduce this involvement, but may increase the promoter's exposure to other risks. The following principles should be used in determining the number of contracts.

(1)   Each contract must be of manageable size and consist of elements that the contractor can control.
(2)   The contract size must be within the capacity of sufficient contractors to allow competitive tendering. Occasionally, only one contractor may be capable of undertaking certain types of work, and competitive tendering would not be possible. Such commitment should be minimised.
(3)   The time constraints of the work may require parallel activities, and this may mean that capacity restrictions necessitate separate contracts rather than a single contract.

In setting the number of contracts, it is useful to list the elements of work and the contract phases from conceptual design to test and commission. These can then be considered in a number of different combinations to decide the minimum number of manageable contracts.

### Tender stages

There are two stages at which the promoter and the project manager can control the selection of contractors: first, before the issue of tender documents, and second, during tender analysis/contract negotiation and placement up to contract award. Both evaluations are important, but they have different objectives:

❑   *pre-tender* means to ensure that all contractors who bid are reputable, acceptable to the owner, capable of undertaking the type of work and the value of the contract, and are competent to manage that type of contract;
❑   *pre-contract* means to ensure that the contractor has fully understood the contract, that his bid is realistic, and that his proposed resources are adequate (particularly in terms of construction plant and key personnel).

## 12.4  Contractor pre-qualification

As a general principle, the promoter should always operate a pre-qualification procedure for project contractors. The choice is whether to adopt a full pre-qualification procedure specific to each contract, or whether to develop standing lists of suitably qualified contractors for various sizes of contract and types of work. Pre-qualification is never a totally objective matter, but subjective judgements should be kept to the minimum wherever possible.

A full pre-qualification procedure may include:

❑ a press announcement requiring responses from interested firms, or a direct approach to known acceptable firms;
❑ the issue by the promoter of brief contract descriptions including value, duration and special requirements;
❑ the provision of information by the contractor, including an affirmation of willingness to tender, details of similar work undertaken, financial data on the number and value of current contracts, turnover, financial security, banking institutions, and the management structure to be provided with the names and experience of key personnel;
❑ discussions with contractor's key personnel;
❑ discussions with other promoters who have experience of the contractor.

The evaluation may be done qualitatively, for example, by a short written assessment from a member of the project manager's staff to narrow down the number of suitable contractors.

After the potential bidders have been interviewed and evaluated, the project manager should recommend a bid list for the promoter's approval. This should be a formal document that provides a full audit trail for the selection process. Specifically, it should discuss the reasons for the inclusion or exclusion of each contractor considered, and confirm that all of those selected for the bid list are technically and financially capable of completing the works satisfactorily.

It is the usual practice to pre-qualify about 1.5 times the number of contractors to be included at tender. This is often achieved by a combination of quantitative and qualitative methods, as no standard procedure exists.

*EU procurement directives*

All companies within the European Union are subject to obligations, under the Treaty of Rome, not to discriminate on the grounds of nationality against contractors or suppliers of goods and/or services from elsewhere in the Union. Some organisations, primarily national and local government and state-owned or privatised organisations, are also bound by the EU Procurement Directives. These Directives require that all procurement of works, equipment and certain categories of services above minimum defined values must be procured in accordance with procedures laid down by the Directives. These procedures involve:

❑ publication of pre-information on procurement intentions in the Official Journal of the EU;
❑ notification of individual invitations to tender in the Official Journal;
❑ prescribed award procedures;
❑ stated criteria for rejection of unsuitable candidates;
❑ permitted proofs of economic, financial and technical standing;
❑ non-discriminatory selection of tenderers;
❑ publication of prescribed contract award criteria;
❑ publication of award details;
❑ the debriefing of unsuccessful tenderers.

## 12.5  Contract documents

For a small amount of work, it can be sufficient for a contract to consist of a drawing and an exchange of letters, or a simple 'purchase order' issued by the promoter to the contractor and acknowledged by the contractor. The drawing can show the location and the amount of work. The materials can be specified on the drawing, and the completion date, price and other terms stated in a letter or in the order.

For the supply of equipment, a specification describing the promoter's requirements in detail will generally be necessary, since normally no drawing or series of drawings can contain enough information.

If the contract is the result of a series of written or verbal interchanges or negotiations between the parties, the final contract must state all that has been agreed and replace all previous communications so as to leave no doubt as to what the terms of the contract are.

To avoid doubt on all but the smallest of projects, the practice has evolved of setting out the terms of the contract in standard sets of documents. These can be lengthy, but will often be virtually identical for many contracts and so do not have to be prepared anew each time. The set of documents traditionally used in UK engineering contracts covers:

❏ agreement;
❏ general and special conditions of contract;
❏ specification;
❏ drawings;
❏ schedules.

A civil engineering or building contract would also contain a separate schedule of rates, or bill of quantities.

The basis of the contract is established by the tender documents. The tender documents must provide a common basis on which contractors can bid, and against which their bids are assessed. The issues between the parties are then clarified, and settled by subsequent negotiation between the parties prior to the award of the contract. The precise terms of the contract are then defined by the contract documents.

The contract should be concise, unambiguous and give a clear picture of the division of responsibilities and legal obligations between the parties. Risks should be identified and clearly allocated.

## 12.6 Tender review

For larger projects, a dedicated tender review team will be required by the promoter. Practice varies from industry to industry and organisation to organisation, but this might comprise a core of two or three people supported by specialists reviewing particular aspects of the different tenders. A typical core team would include the project manager, the lead designer or process specialist, and a representative of the user department, with specialist support from other design specialisations, a contractual/legal expert and a quality assurance (QA) specialist. The tender review criteria will depend upon the chosen contract strategy.

*Lump sum:* with this type of contract, the objective of the review is to find the tenderer who offers the lowest price and best programme, and yet meets the specification in terms of scope, quality, operability and economic maintenance. It will need to concentrate on identifying areas where tenders do not comply with the promoter's requirements. If the contractor is to have design responsibility, design capability will also be reviewed if this has not been dealt with at the pre-qualification stage.
*Reimbursable cost:* the objective with this type of contract is to find the

tenderer who has appropriate design and management skills, as well as a project execution capability that gives the promoter confidence that it will meet the project requirements. The tenderer must also understand the promoter's requirements and offer a collaborative management team who are able to work with the project manager.

*Quantities/rates-based and other contracts:* will fall somewhere between these two extremes in terms of the promoter's risks, and the review team must make a judgement on which tender will provide the best value for money.

## 12.7  Tender evaluation

The project manager will open the tenders from the various contractors on a given date and at a given time. A systematic evaluation of the tenders would include an examination of:

❑ compliance with the contractual terms and conditions (and any qualifications made by contractors);
❑ technical correctness of tender;
❑ design or other advantages offered by the different contractors;
❑ correctness of bid prices (e.g. if errors are detected in multiplying rates by quantity);
❑ screening of bids for detailed analysis;
❑ pre-award meetings/negotiations (optional, but usually essential);
❑ selection of the best bid and recommendation to the promoter for contract award.

*Bid conditioning* is a term which is sometimes used to define a process of reviewing all tenders in order to be able to compare like with like. Some organisations take a rigid stance and reject all tenders that fail to conform totally with the inquiry requirements. Others accept that tenderers may offer alternatives of genuine benefit to both parties, and hence consider all submissions. This second approach is normal within the equipment/process industries, although it is rather less common in the civil engineering and building industries, for obvious reasons. In that event it is usual to require tenderers to submit a 'conforming' bid as well as variants to allow comparisons to be made. Where direct comparison cannot be made, exceptions must be carefully noted.

A misconceived tender based on an error or misunderstanding by the tenderer should be easy to identify during tender evaluation. Exceptionally low bids are automatically rejected by some promoters as they suggest that the contractor has made basic errors, or is desperate to obtain work at any price. However, in some circumstances a very low bid will be a strategy for 'buying' the job for the contractor's own valid commercial reasons. In the evaluation, an attempt must be made to discover the contractor's philosophy. If the promoter is to accept a low price, then it is necessary to ensure that the project will be completed for this price. The pre-award meeting/negotiation is the ideal time to discover the motives of the contractor.

### Contract award recommendations

It is essential that the review team should put together a formal recommendation for the award of the contract. The report should:

- ❑ explain the background;
- ❑ summarise the recommendations;
- ❑ describe bid opening and the initial position;
- ❑ describe any bid conditioning process;
- ❑ give reasons for rejection;
- ❑ identify the tender recommended;
- ❑ summarise reasons for recommendation;
- ❑ set out the cost, time and other implications for the project.

## 12.8  Typical promoter procedure

Sometimes the tender packages are prepared by an organisation employed by the promoter, or by a consultant engineer or management contractor who is responsible for the design. In either case the procedures to be followed are similar, although the titles of the departments and individuals concerned may differ between organisations. In the case of large projects, such as those implemented offshore, the level of project management required often makes the services of a management contractor extremely cost-effective. For the sake of conformity, the procedures will be described as if a management contractor has been employed by the promoter to coordinate the project.

The project manager controls the preparation of the tender package, possibly as the head of a department with duties solely relating to the preparation of contracts. The main purpose of the contracts department is to check:

❏ that no gaps or overlaps exist between the individual subject contracts;
❏ that all the particular requirements of the owner are included and covered in the subject contracts.

In particular, the contracts department will liaise and coordinate its work with the other departments with regard to matters such as:

❏ engineering/design;
❏ quality assurance/quality control (QA/QC);
❏ materials/procurement;
❏ construction;
❏ law and insurance;
❏ planning and scheduling;
❏ costs and estimates;
❏ accounts and payment;
❏ performance;
❏ planning and other approvals;
❏ licencing;
❏ safety/environmental considerations;
❏ 'buildability/operability'.

These departments in turn should liaise with their counterparts in the promoter organisation.

The individual tender packages are usually developed from pro-forma/standard documents developed as standards for the project by a selection of alternatives and/or amendments.

Draft formats of the individual subject contracts should be circulated to each department for review, allowing sufficient time to incorporate amendments within the tender package. A standard procedure should be adopted to report comments and amendments to the contracts department.

The final draft of the tender package is then sent to the promoter for formal approval prior to the issue of tenders. The contracts department will maintain a register of tender documents, including information such as:

❏ reference numbers;
❏ description of requirements;
❏ date of issue to the tendering contractors;
❏ tendering contractors;
❏ contractors' questions, and answers;
❏ date of return of tenders, and the date to which tenders remain open.

The procedures and pro-forma documents used by the contracts department in preparing tender packages should be set out in a document (sometimes referred to as a works plan).

### Invitation to tender

This should be in the form of a letter to tenderers written on behalf of the promoter. The letter simply invites the contractor to submit a tender for the performance of certain work. The letter lists the tender documents attached, and requests the contractor to acknowledge receipt of the documents and their willingness to submit a tender. A 'form of acknowledgement' is usually included with the letter so that replies from all tenderers are set out in a standard way.

### Instructions to tendering contractors

Instructions to tendering contractors inform the tenderers what is specifically required of them in their tender, and usually comprise:

- ❏ tendering procedures;
- ❏ commercial requirements;
- ❏ information to be submitted with the tender.

It will normally be made clear to tendering contractors that the tender submission should be based on the scope of works described in the contract documents, and that any permissible alternative tenders are to be submitted with a conforming tender as set out in the original tender documents so that the promoter may compare the two. Where the promoter will not consider alternative tenders, this should be expressly stated in the instructions to tendering contractors.

### Conditions of contract (articles of agreement)

The conditions of contract can be prepared by the contracts/legal department in consultation with the project manager and insurance/finance/shipping disciplines, etc. For an international contract, legal and other experts in several different countries may need to be consulted. Standard conditions are prepared for the whole project, and must be agreed with the promoter.

The conditions of contract are often based on an industrial standard or the promoter's standard. In either case it is likely that modifications will have to be made to suit the unique requirements of each project, the structure of the contract package, and the philosophies of the contracting procedures (see above). The standard conditions can be modified as necessary for individual contract packages.

The promoter and the contractor will usually be the parties to the contract, unless another arrangement is required for reasons such as project finance. A project manager be may appointed by the promoter, via a separate contract of employment, to monitor the performance of the works.

### Brief description of the works

The brief description of the works explains the overall requirements and parameters of the works to be performed. The description should be drafted with the aim of creating a broad appreciation of the works. The overall dimensions of the project should be stated, and technical links between the elements of the contract package should be described.

The description should be kept brief and simple and should not repeat detail already covered by the drawings, specification, programme or any other work element of the contract package. However, it should define any work that is not covered elsewhere in the contract package. The interface with other contract packages should be described without reference to the actual work contained therein.

This section may also be used to detail such things as the facilities that are required to be provided on site by the contractor.

### Programme for the works

For larger projects, the programme is usually prepared by the planning and scheduling department liaising with the contracts department. It should contain all key dates for the particular contract, including:

- award of contract;
- start of the works;
- completion of the works;
- intermediate key dates for particular elements of the work, where such elements are required to interface with work outside the scope of the subject contract;
- critical dates within the contract.

### Indexes of drawings and specifications

Indexes of drawings and specifications are usually prepared by the engineering department. They should be correctly numbered to identify the latest numbers and revisions of drawings and specifications contained in the tender packages. The indexing system will then usually apply throughout the life of the project.

The format, but not the contents, of the indexes should be prepared as a standard by the contracts department in consultation with the engineering department.

### Drawings and specifications

The drawings and specifications are prepared by the engineering or quality control and assurance departments. Generally, two copies of each drawing and specification are included in the tender package for each bidding contractor.

The terminology of the specifications must be consistent with that of the contract documents. The contracts department should review all specifications before issue to ensure that the terminology is consistent. The contracts department should check the individual parcels of drawings and specifications against the indexes in the tender documents. The specifications and drawings issued with tenders (i.e. those listed in the index of specifications and drawings) should be given an identifying revision number.

### Promoter-provided items

Where items are to be supplied by the promoter, they must be listed. This would normally be done by the engineering department in liaison with the contracts department. The following items are usually considered:

- descriptions of the items provided;
- delivery periods;
- specific storage requirements;
- explanation of any markings;
- details of returnable packaging.

Practical delivery periods should be stated against each item. It should be noted that such periods become contractual commitments and should be strictly adhered to.

### Contract coordination procedures

The contract coordination procedures describe the administrative requirements for the implementation of the contract. The procedures explain the day-to-day duties and responsibilities of the site management team and the contractor's site team. The procedures also detail the lines of communication.

Guideline procedures detailing the implementation of technical requirements can also be included, but this is generally inadvisable. If they are included, a check must be made for any duplications or contradictions with the specifications.

In compiling the coordination procedures, the construction department will need to work closely with the various disciplines to obtain their requirements for the implementation of their specific responsibilities.

## Form of tender

The form for a tender is prepared by the contracts department. Practice varies widely between different industries, but a typical form might be as shown in Table 12.1.

**Table 12.1** Preparation of a tender package.

| Section | Pro-forma used as standard | Individual sections to be pepared |
| --- | --- | --- |
| Invitation to tender | X | |
| Instructions to tendering contractors | X | |
| Conditions of contract | X | |
| Brief description of the works | X | X |
| Programme for the works | | X |
| Index of specifications | | X |
| Specifications | (format) | X |
| Index of drawings | (format) | X |
| Drawings | (format) | X |
| Items provided by owner | (format) | X |
| Contract coordination procedures | X | |
| Draft schedules | (format) | X |
| Form of tender | X | |
| Form of agreement | X | |

### Review of the tender package

Following the assembly of the tender package, a copy is sent to the department managers of the management contractor for review and comments. Following the reviews, any comments should be incorporated where necessary. Reasons should be given by the contracts department for not incorporating any comment. At least two copies of the reviewed tender package should be sent to the promoter for review and comments.

### Collation and issue of tender packages

The final collation of the tender packages and the issue to the individual tendering contractors should be in accordance with the contracting schedule and is normally the responsibility of the contracts department.

### Queries from tendering contractors

All queries from the tendering contractors during the tender period should be answered by the contracts manager. Any other contact between the tendering contractors and the promoter or other members of the staff during this period is to be discouraged.

## Further reading

Boyce, T. (1992) *Successful Contract Administration*, Thomas Telford, London.
Turner, J.R. (1995) *The Commercial Project Manager*, McGraw Hill, London.

# Chapter 13

# Project Organisation Design/Structure

Earlier in this book, the evolution and development of projects was examined. No project operates in isolation, and there is always interaction with the promoter's teams and external and internal specialists and suppliers. The project needs to develop an appropriate organisation to function within the project environment and with external parties in order to fulfil its objectives. This chapter is concerned with the manner in which projects and interacting organisations are structured.

## 13.1 Organisations

Engineering projects consume large amounts of resources, are often complex and vary in scale. It is therefore essential to adopt a systematic approach to managing these activities. At the heart of all managerial activity is the creation of an organisation to execute the managerial objectives, and to establish how the people within it relate to each other. Organisation is about creating a control and communication system that allows management to achieve its objectives. It puts in place a structure that defines roles and hierarchies, and communication, coordination and control mechanisms.

Larger and older organisations tend towards bureaucracy; that is, they operate extensive systems of rules and regulations to manage the organisation. This approach is adopted to achieve control and stability. Smaller organisations tend not to have extensive rules and regulations and are more flexible in their approach. The more bureaucratic organisations tend to be less flexible, more predictable and better suited to stable rather than dynamic conditions. The more flexible, organic organisations respond faster and adapt more easily in changing and dynamic situations. Interacting with either type of organisation will have an impact on how the project organisation is developed. The design of the project organisation has to serve two masters, one involves flexibility, speed and dynamism, the other processes, planning and procedures. This should allow it to interact with existing organisations, but also fulfil its own goals.

## 13.2 Building blocks of organisations

There is no universal system of organisation that suits all circumstances. Modern attitudes to organisation design refer to the 'contingency approach', which suggests that the most appropriate form of organisation is contingent on the influences on the organisation. Project organisations are temporary in nature, but often have to interact with permanent external and internal organisations. The manner in which these interactions take place will have an influence on what the project organisation can achieve. By understanding the influences on the design of interacting and sponsoring organisations, it is possible to orientate the project organisation accordingly.

Every organisation, large or small, is influenced by its approach to the system of authority. This is commonly referred to as the hierarchy of the organisation. Hierarchy relates to the number of levels of authority and control, and where decisions are made. The hierarchy is influenced by the senior management's attitude towards control. The closer the control or supervision of subordinates, the greater the number of levels in the organisation. The classic approach to organisation design resulted in a pyramid-shaped hierarchy, that is best illustrated in traditional organisation charts. The hierarchy in an organisation is a function of the senior manager's attitude to control.

The next element in organisation design is how the roles of the members of the organisation are defined. A rigid system of role definition and responsibility creates a sense of stability in that there is a place for everyone and everyone knows their place. Bureaucracy was founded on the principles of having this sense of order. However, this approach may stifle flexibility, creativity and prompt decision making. More recent approaches to role definition tend to set more flexible parameters for the participants of the organisation. Roles are defined to allow for more autonomy and scope for individual innovation. This approach resembles the organic organisation form. The less rigid approach to role definition allows hierarchies to be broken down, and places more faith in the people in organisations.

A balanced hierarchy and sensible role definition do not necessarily create a good organisation. The design of the organisation should also facilitate coordination and communication. An organisation that does not coordinate its activities or communicate efficiently will face difficulties. Communication should be both vertical and horizontal. That is to say, communication should occur up and down hierarchies, as well as between units at a horizontal level. The emphasis on the need for communication should be built into the early thinking in the organisation design. This pre-planned approach is becoming more important as organisations become more temporary and use more external sources for achieving goals. The design of the organisation should also serve a coordinating function. Traditionally, organisation design has tended to use the division of work as a means of creating structures. This is a sound approach where projects and products are simple, and high levels of coordination are not required. In an organisation that has complex projects and products, an organisation design to help facilitate coordination is essential. There is a move towards creating multi-disciplinary teams and cross-functional groups that interact more freely and are not constrained by specialist boundaries. In these circumstances, good communication and coordination have to be designed into the organisation.

Most of the design dimensions we have considered relate to the internal perspective of the organisation, but the most important influence on the organisation is how it relates externally. The customers, markets, regulators, shareholders, government and so on can ensure success or create failure. Burns and Stalker (1986) in work on organisations, showed the impact of structure on success or failure in turbulent environments. Peters and Waterman (1982) in their study of 'excellent companies', also highlighted the importance of external orientation for organisations. It is difficult to identify individual factors that make organisation design more responsive to external conditions, but there are certain characteristics that should be present. Decision making should quick and decisive; there should not be a situation were there is 'paralysis by analysis'. The structure should allow the company to be responsive to changing markets and customer needs. Role definitions should have loose and tight properties, i.e. loose to empower staff to take decisions and be innovative, and tight to ensure that the organisation's goals remain a key focus. External orientation in organisation design cannot be built into the structure, but it develops as a result of the need to respond to external circumstances. The role of senior management in creating a responsive structure becomes paramount.

How does the understanding of these building blocks of organisation design help in managing projects? They assist by creating a structure for project organisations, and helping to improve our understanding of the interactions between projects and permanent organisations.

Organisation design is contingent upon the influences on the firm; hence, there is no right answer. Probably the best advice on organisation design was given by Peters and Waterman (1982) when they suggested that structure should be simple, lean and easily understood by those who work in it.

## 13.3 Organisation types

Although we have considered the dimensions of organisation design, there are a number of generic organisation designs that exist. These organisation forms have been developed over time and are present in engineering organisations that operate today. We are concerned about the factors that underlie their creation, their strengths and weaknesses, and how they might influence projects that interact with them.

### Functional

In engineering, and in many other fields, jobs have become more specialised. The growth in specialisation and expertise has had an influence on the way organisations develop. Specialisation is an extension of the principles of division of work, and specialists tend to group together and form teams, sections, units and departments. In its simplest form, groupings of specialists give rise to the functional structure (Figure 13.1). Functional organisations arise out of the principle of division of work, build on specialist skills and dominate most organisations through specialists. Expertise and information are contained within each specialist function, group or department. Projects operating within a functional environment are reliant on the cooperation of specialist

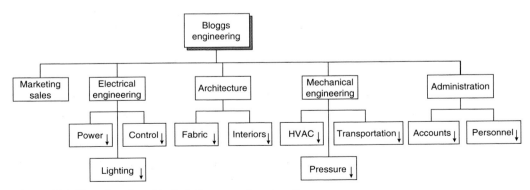

**Figure 13.1** Functional multi-disciplinary engineering firm.

functions and communication with other departments. Different specialist departments often guard their own expertise, have their own specialist objectives and are not good at integrating with other experts. They may lack a common interest, and also lack of understanding of other specialists, and consequently communication and decision making suffer. It is this separation of interests that pose problems to projects. Individual specialist interests are promoted ahead of project goals, and the potential to disrupt the project cycle is more likely. Decision making is based on ensuring functional performance. Projects operating in a functional environment require project managers to concentrate on integrating activities and communication with specialist departments.

### The division

Some companies try to overcome the particular problems of functional organisations by creating organisational structures that have more focus and facilitate communication between specialists. Organisations of this type design their structure by focusing on the specialist nature of the work or project rather than on individual expertise. This form of structure is known as the divisional structure (Figure 13.2). The aim of the divisional form is to replicate specialist characteristics, but to focus on a final product through its specialist type, size, location, customer, etc. Participants at every stage of the product process are brought together to overcome the specialist niche mentality, to improve communication, and to identify with the end product. Projects are easier to facilitate within a division, as all the components within the organisation are geared to meeting the division's goals. Divisional organisations are predominantly found in large companies. Moreover, the entire focus of activity is on the division, and decision making is based on divisional

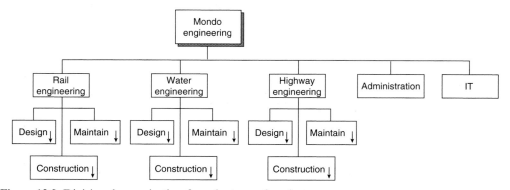

**Figure 13.2** Divisional organisation form by type of project.

requirements. Projects are vulnerable where they operate between the boundaries of divisions, since inter-division rivalry, poor communication and inefficient information exchange may exist. Projects developing in the divisional environment require project managers to be aware of divisional and corporate priorities and how they may impact on projects.

### The matrix

The third predominant organisation form that has arisen developed out of the problems encountered with integrating projects into existing organisations. Projects, by their very nature, are temporary and are difficult to integrate with permanent organisation design. This problem is further exacerbated in organisations that tend to manage largely by projects, which usually involves a balancing act between permanent specialist functions and temporary project structures. The matrix structure (Figure 13.3) attempts to resolve this problem by imposing a temporary project structure across the permanent specialisms. The idea is to move groups of specialists to different projects as they are needed. Projects fall under the control of project managers, resulting in the situation that specialists are responsible to two managers. Working in a matrix requires careful role definition and a clear authority framework in order to prevent conflict. The personal skills of the project manager are important in balancing the requirements of the project with the specialist

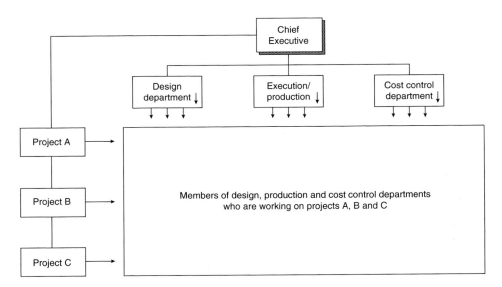

**Figure 13.3** Matrix organisation structure.

functions of the organisation. These problems become more prominent with planning and resourcing constraints within the firm. The matrix has been used successfully as a structure to manage internal projects, as it is geared towards accommodating projects.

### The hypertext organisation

A further development of the matrix organisation has resulted in the hypertext organisation. This term was suggested by Nonaka and Tekeuchi (1995) in their book *The Knowledge-Creating Company*. This organisational form views bureaucracy and project teams as complementary rather than mutually exclusive. The metaphor that is used for such a structure comes from hypertext. Hypertext consists of multiple layers of text, while conventional approaches consist of one layer of text. Under hypertext, each text is stored in different layers or files. It is possible not only to read through text, but also to delve into different layers of detail and background material, thereby creating a different context. The hypertext organisation (Figure 13.4) is made up of interconnected layers. The central layer is the main business systems layer in which routine operations are carried out, and it may have a conventional hierarchy. The top layer is the project layer, where multiple project teams are engaged in new projects. The project team members are brought together from the different units operating in the business system, and are assigned exclusively to the project teams until projects are completed. The third layer does not exist as an organisational entity, but is embedded in organisational culture, corporate vision, technology, etc. It

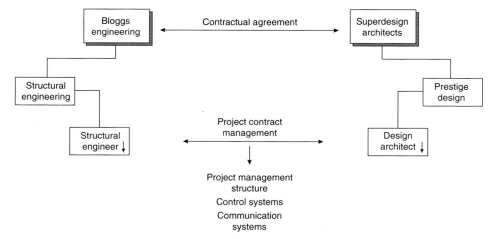

**Figure 13.4** Collaborative contractual relationships.

is the layer that helps contextualise the other layers and provides the basis from which the organisation can learn and develop. The third layer also helps to focus the organisation externally, creating a more open system approach. The hypertext form of organisation is driven by the deadlines set by projects, and resources are concentrated to achieve this. A team member in a hypertext organisation belongs, or reports, to only one part of the structure at any one time, either the project or the business system. The vision, culture, technology and learning layers ensure that what is developed in other layers is communicated around the organisation. The key characteristic of the hypertext organisation is its ability to shift people in and out of context to one another.

*Networks*

There is a trend in engineering companies to concentrate on their core businesses or activities in which they have the greatest competency. Some projects cannot be sourced entirely from internal resources, and external resource inputs are needed. How are externally sourced projects structured? The elements of designing an organisation, such as role definition and hierarchy, still remain an essential part of structuring an external project. The structural form that emerges is a network (Figure 13.5) which places the project manager at the centre. Hierarchical positions and roles are defined by contracts. The project manager is at the centre of

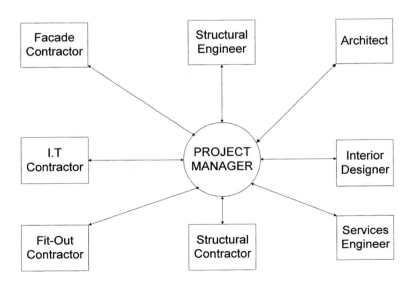

**Figure 13.5** Network organisation.

a nexus of contracts. These formal relationships allow the project manager to exercise authority within the project. The project is initiated through an external project champion, while the project is achieved through external specialists. The role of the project manager is to plan, monitor and control these relationships. The key to operating under this structural form is the power that is vested in the project manager. This power will largely be derived from the nature of the contracts set up between the project sponsor and external specialists. As the trend towards specialisation increases, the project manager will become more and more involved in managing projects that arise from networks of companies coming together for a project. Charles Handy, in his book *The Empty Raincoat – Making Sense of the Future* (1994), suggests that as more collaborative forms of management come into use, managers will in essence be managing federations of companies that are networks.

### Virtual organisations

The past decade has seen profound shifts in our understanding of organisations and what it takes to be a member of an organisation. The contracts and spatial relationships between the organisation and its members have been redefined. A major redefinition of the open-ended employment contract has taken place, with greater use of temporary flexible arrangements. New forms of working involve different combinations of contractual and locational variables. This creates a working environment that is more fluid and flexible, and without permanence. A new form of decentralised organisation, the virtual organisation, has now emerged. The virtual organisation is difficult to characterise in terms of work patterns, organisation structure, boundaries and physical form. They are temporary organisations that produce results without having form in the sense of traditional organisations. The virtual organisation defines roles in terms of the task of the moment, rather than the role anchored by the organisation and a codified job description. Time, space, the tasks and shifting group membership are the primary definers of responsibility. People are moved in and out of the organisation as and when they are needed. The virtual organisation is taking network organisations a stage further. They are networks of people and activities brought together to fulfil particular tasks without the bounds of traditional organisational form. Virtual organisations maximise the use of people and knowledge rather than investing in the costs of permanent organisations. The use of virtual organisations on engineering projects will increase as we use more specialists and experts on a shorter time, task and role basis. The virtual organisation is an exchange network, and will continue to develop as an organisational form as we improve communication technology. The primary constraint on virtual organisations is that it does not conform to the traditional approaches to management.

## 13.4  Internal and external projects

Many lessons on how we structure our project organisations can be taken from understanding general organisation structure. A more important aspect of understanding structure is its interaction with projects. All projects have to interact with organisation structure. Internal projects, which operate within the boundaries of the organisation, have to accommodate structural influences. Internal projects are characterised by their reliance on the existing structure of the organisation. Resources for projects are drawn from existing departments, divisions and business units. Existing operational systems, hierarchies and cultural influences are also drawn into the project. The ability to structure an internal project, without interference from the rest of the organisation, is dependent on the power and authority vested in the project. A powerful internal project champion may help the project overcome conflicts with the existing hierarchy. Conflicts with existing departments will hamper projects. In order to facilitate and execute internal projects successfully, organisations have to create reporting relationships, systems and roles that accommodate projects within their boundaries.

External projects differ from internal projects in that they operate outside the boundaries of the organisation or have a significant external influence on them. When a project operates outside the boundaries of the parent organisation, the ability to control it is reduced. The project becomes more susceptible to a variety of external influences. Issues such as politics, economics, societal values, competition, markets and technology take on a more important role in influencing the success or failure of the project. The project manager also has to ensure that the needs of the project sponsor are balanced with the external challenges a project faces. External projects may also be characterised by numerous participants that are from outside the sponsoring organisation. The project manager has to ensure that contracted roles and responsibilities are clearly defined, with a sound control system in place. The structure and culture of such a project team is determined by the project manager, as there may not be the overriding influence of one organisation. While this offers flexibility in the structure of the project, it also means that the learning curve and the group dynamics of the project team take longer to achieve. Projects with an external orientation have to respond more quickly to changes in the environment. The manner in which the team is structured should be flatter and more responsive, taking on the characteristics of an organic organisation. Resource conflicts and associated problems also exist on external projects. The main task in structuring an external project is to ensure that the orientation of the project will respond to external influences.

## 13.5  The human side of structure

Organisation structure in itself is not the end, but the means, of achieving organisation and project goals. People have to interact within these structures to achieve success.

### Leadership

At the centre of the structural arrangements are leaders. The role of the leader in setting the style of management and the manner in which the structure is operated is crucial. There are the two extremes of leadership style, the centralised approach and the decentralised approach. There are supporters of both these views, but it is also important to consider which approach is suitable to the task in hand, and also whether the structure is suited to the task. The traditional approach to leadership was to centralise decision making and exercise control through a pyramid-shaped hierarchy. This is regarded as a top-down approach to management. In clearly structured task environments, the centralised approach works well, as there is a need for clarity of direction. More recent views with regard to control and leadership tend towards decentralised and participative decision making. Organisations increasingly rely on specialists that come from all levels within the hierarchy, and it is their knowledge and expertise that influences decision making. A more participative environment is better suited to dealing with complex problems where more ideas and opinions are needed. Decentralised leadership transfers the responsibility of leading to the individuals or groups involved in the projects. The role of the leader is important in that the leader's style of management can set the tone and culture of the organisation and the way it performs.

### Project teams and empowerment

Engineering firms have also made increasing use of dedicated project teams. Establishing project teams within existing structural arrangements is always fraught with problems, and a clear relationship needs to be established between the project team and the rest of the organisation. Role definitions, reporting relationships, power and authority need to be addressed. Equally important is the development of the project team. Projects are temporary, and hence project teams are temporary, but they have to achieve their objectives without the luxury of a second chance or prolonged learning curves. Project teams need to develop a sense of unity, purpose and identity fairly quickly in order to become fully functional. Prior to the project team becoming directly involved in the project, it is common that some form of team-building exercise takes place. This allows the project team to bond, and these measures may also be backed up by locating the team together and using other symbols of a distinct project identity. The extent to which a project team is successful is not entirely about human interaction; it is also about the extent to which the team is empowered to undertake its task. Empowering a project team is about transferring responsibility and authority to the team. The key to empowerment is ensuring that those within the team feel a sense of ownership and control over the project. The greater the sense of project ownership, the more the individuals within the team are likely to work towards succeeding on the project. The transfer of the control of the project to the project team creates problems with the traditional structure, but these have to be overcome in order for project teams to succeed.

The people within a structure have a great deal of influence on the extent to which the organisation can succeed. They are the source through which the organisation and project goals are transformed into reality.

## 13.6  Structure in collaborative relationships

The manner in which projects are carried out has changed over time. Early management thinking operated with the view that one company conducted all the activity required to produce a product. The view today is that a firm should be good at what it does, but it cannot be good at everything. This focus on core competencies has led to the situation where collaboration between firms is more common. Collaboration takes many forms, such as strategic alliances, joint ventures, partnerships, joint ownership, etc. These forms of working have an influence on projects and on existing organisations. They influence hierarchy, roles, control and goal setting.

At the heart of all collaborative relationships is partnership. It is safe to assume that considerable thought would have gone into the use of a collaborative relationship and the selection of the right partner. This is the start of the relationship, and there are a number of pitfalls that may damage it. One of the first issues to consider is how the collaboration is structured and managed. Is the collaboration going to take on a conventional structural form, a network or a temporary project team? Each system will have merits and negative factors. Consideration will also have to be given to how integration is going to take place with the parent firms. Another factor that has an influence on collaborative projects is who leads the venture and from which firm does this person come? The question of bias towards one partner or the other may arise as a result of the senior project manager's decision. It is also possible to select an 'unbiased' external manager to head the collaboration, but then there is the possibility that there will be no political support from parent organisations should the project start to fail. The leadership arrangement will ultimately be a compromise between the collaborating companies.

The control exercised by parent companies will also influence the structure of the collaboration. A parent company that has a bureaucratic approach will expect the collaboration to conform to these ideals, while one with a more organic approach will be more flexible. Ideally, the collaboration will be empowered to achieve the targets that have been set for it with minimal parent company interference. This is an idealistic situation; reality suggests that because resources are involved, the parent companies will attempt to influence the collaboration. In collaborative ventures, the more powerful parent company is also likely to exercise undue influence. It is essential not to forget the role of the parent company as the initiator of the collaborative project. They have to be kept appraised of how the venture is progressing. Communication protocols should be established to ensure that issues relating to the collaboration get to the relevant levels within the parent company's management structure. The parent company is an important stakeholder in collaborative projects.

Alliances, partnerships and joint ventures involve people from two or more groups of companies. Successful partnerships are not just about the contracts, but also about the relationships. A good relationship goes a long way to making the project successful. Further discussions on collaboration can be found in Chapters 15 and 16.

## 13.7  Structure in the international context

Improvements in transport, communication and technology have made the market place more global. Companies are faced with the prospect of undertaking projects that are away from their home country in order to remain competitive. Operating a project that is in an environment that the company is not familiar with throws up new challenges. The most tried and tested organisations are often required to change the manner in which they work when faced with different cultures.

The overriding influence on organisations is the impact of a change in culture. The world is a diverse place, and the differences in culture may become apparent when international projects are undertaken. The word 'culture' means the collective symbols, values, rituals, heroes and practices that are common to a particular society. These concepts may not be common to all the people within the society, but they are a fairly good generic collective expression of their culture. How does culture influence the way in which we manage? Gert Hofstede's seminal book on work-related values (Hofstede, 1994) gives us some indication of how different people view their work. He came up with four key dimensions of national culture which influence attitudes to work:

❏ individualism versus collectivism;
❏ large versus small power distances;
❏ strong versus weak uncertainty avoidance;
❏ masculinity versus femininity.

In individualist cultures people look after their own interests, while in collectivist cultures they remain unquestioningly loyal to larger groups. Power distance represents the extent to which less powerful people in a culture accept and expect power to be distributed unequally. Uncertainty avoidance represents the extent to which people find difficulty in coping with unstructured and uncertain situations, and prefer stricter rule-based situations. They are less tolerant of deviance and prefer certainty. In masculine cultures, the society is dominated by males, who are expected to be assertive and competitive, while in feminine cultures, the environment is less confrontational and relies on relationship building. These combinations and their variants appear in every society. Culture will have an influence on the effectiveness of the management approach and structure being adopted. It is therefore necessary to consider how the structure and management of a project should be adapted to take account of the way these dimensions influence working practices in a particular country or culture. Many international projects have run into trouble because of a lack of awareness of cultural variations. Engineering projects in an international context require work to be carried out in other countries, and involve the need to collaborate with international partners. From a managerial perspective, the potential impact of the local culture on an international project cannot be neglected.

Engineering projects may transfer technology to other countries, but a great deal of what is achieved requires local input. Planning permits, approval and other forms of permission have to be achieved at a local level. It is common practice in certain circumstances to have a local facilitator on projects. The facilitator becomes part of the structure, and usually works as a consultant to the project manager. Careful thought should go into the integration of the facilitator into the structure and how this role is defined. It is the role of this person to ensure the smooth integration of the project into the local situation. The expectation is that the facilitator will help with the necessary local requirements for the project. This individual should also be a rich source of intelligence on local culture, politics and working practices. Ultimately, the facilitator must be a good diplomat with a network of contacts.

## 13.8 Summary

The management of organisations requires a systematic approach in order to create a sense of order and control. The lessons from this approach that can be adopted for projects are:

- ❑ clear definitions of the roles of the participants in the project;
- ❑ an established and clear relationship between the project and its sponsoring parent company;
- ❑ an external orientation in setting up the project management structure;
- ❑ the realisation that more projects will require collaborative relationships;
- ❑ people and the way they are led is the key to making a structure work;
- ❑ on international projects, an understanding of cultural diversity is essential for better management.

## References

Burns, T. and Stalker, G.M. (1986) *The Management of Innovation*, 2nd edn, Tavistock, London.

Handy, C.B. (1994) *The Empty Raincoat – Making Sense of the Future*, Hutchinson, London.

Hofstede, G. (1994) *Cultures and Organisations. Software of the Mind: Intercultural Cooperation and its Importance for Survival*, HarperCollins, London.

Nonoka, I. and Takeuchi, H. (1995) *The Knowledge-Creating Company*, Oxford University Press, Oxford, New York.

Peters, T.J. and Waterman, R. (1982) *In Search of Excellence: Lessons from America's Best-Run Companies*, Harper & Row, New York, London.

## Further reading

Belbin, R.M. (1993) *Team Roles at Work*, Butterworth, Oxford.

Fielder, F. (1967) *Theory of Leadership Effectiveness*, McGraw-Hill, Maidenhead.

Handy, C.B. (1995) *Beyond Certainty – the Changing Worlds of Organisations*, Hutchinson, London.

Hickson, D.J. (1997) *Exploring Management Across the World*, Penguin, London.

Nohria, N. and Eccles, R. (1994) *Networks and Organisations*, Harvard Business School Press, Boston.

# Chapter 14

# Design Management

This chapter concentrates on the stages of design where the greatest use of design resources usually occurs, from concept through to detailed design. It is also at these stages, particularly concept, that the ability of the designer to determine the characteristics of the project which are deliverable is critical. The ranges within which cost, schedule and quality criteria for the project will fall are determined during the design stages.

Effectively and efficiently managing design is clearly of fundamental importance in the overall project management activity. This chapter will first describe the aspects of creativity in design, then move on to the activities within the design stages, and thereafter discuss the techniques used to manage the design work on a project.

## 14.1 Role of designs

Designers occupy a tremendously important position in a project's life-cycle. The input to the design stages is usually a brief, a statement of requirements, or a performance specification, but the input of the designer to the project does not begin on receipt of this document, in whatever shape or size it arrives. The very creation of the input document is likely to have been moulded significantly by designers, and in some enlightened projects the designer may even have had an input to the business-case development that lead to the initiation of the project in the first place.

Design input to the project does not finish at the time when the implementation of the design solution starts. As the building, facility, infrastructure work, process plant or other type of deliverable engineering project begins to be 'made', design input is needed to guide the implementation people as unforeseen problems arise. Designers will probably be called on during the commissioning stages to assist when the design solution is finally being put into operation. Throughout the operational life of the facility, designers will be called on to create solutions to new requirements not originally envisaged by the project owner, such as new production facilities to cater for increased market demand. When the facility has reached the end of its economic life, designers may well be called upon once again to advise and assist in the process of safely decommissioning and removing the facility.

## 14.2 Understanding design

At the core of the ability to design is some element of creativity. By the nature of the concept, creativity is hard to define and hard to measure. It is often associated with the 'arts' as a generic term to capture all things creative. However, creativity is also fundamental in many other areas of work and life, such as science, engineering and business. Creativity within the context of design is difficult to isolate. There are many areas of industry and commerce where design activities are carried out. These include such diverse areas as creating new clothes fashions, the preparation of food and drink, graphic design, lithography, business strategy, engineering design, industrial design and architecture. The ways in which creativity manifests itself in these different design processes are many and varied, and are generally not well understood.

A number of models have been developed to represent the creative process, and the one which is probably used most frequently is shown in Figure 14.1. This is known as the assess–synthesise–evaluate model. At the assessment stage, the person being creative assesses a number of information inputs, some of which are well known to the creator, such as historical solutions to the particular problem under consideration, and some less well known and more fragmentary in nature. This period is often frustrating, as attempts are made to manipulate the better and less well known elements of information into a solution. There often follows a period of time, in the synthesis stage, when the person working on the problem consciously moves away from seeking a solution, and works on some other activity or problem. At this stage, the input information is being synthesised in the sub-conscious mind. This is usually followed by the 'eureka' moment, when the thinker becomes aware of what seems to be a major breakthrough in finding the solution. At times a fully formulated solution 'appears' in the mind of the creator, although this is often illusory. The person working on the problem is then able to evaluate the solution, or part solution, they have arrived at, and compare it with the problem as they have defined it to themselves in the assessment stage of the process. The ways in which the solution fails to meet the

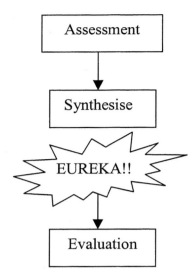

**Figure 14.1** The assess–synthesise–evaluate model of creativity.

needs of the problem are then taken into the assessment stage of the next round of the process, and these three steps are repeated until an evaluation of the solution indicates that the problem has been solved.

There is clearly then some element of unpredictability in the process of creation. The period of synthesis between assessment and eureka can be the wink of an eye, or days or months, and in some cases has even been reported to be years!

There are also a number of different ways of thinking about the progress of the design process itself, as distinct from the creative aspects of the design. These models of the design process suggest that the way design is approached can influence the degree of control that can be exerted over design and designing. Two models of design are commonly used.

(1) *Bottom-up, top-down and meet-in-the-middle.* The bottom-up style combines basic structures and eventually produces a final output. The top-down style conversely works back from a desired final behaviour to sub-behaviours which are linked to components and their structures. The meet-in-the-middle design process style combines both bottom-up and top-down processes. Either method may be used according to the context of each particular aspect of the design.

(2)   *The person-level methods*. With a depth-first strategy, a hypothesised design is thought of first, and then an attempt is made to make it work. Several ideas may be pursued simultaneously until there is convergence to a final solution, or a methodology is used in which designers know intuitively which design to select at each stage in the process.

At a less abstract level design activity can be made more prescriptive, and processes are more and more frequently being captured in design process maps. The common processes modelled include determining functions and their structures, elaborating specifications, searching for solution principles, developing layouts, optimising design forms, and dividing design work into realisable modules. Some element of planning is often incorporated, usually to enable the designers to plan their work better.

Working within these explicit processes should encourage more creativity, since less attention needs to be paid to ensuring that ad-hoc processes are in place to meet the needs and constraints of the project for which the design is being carried out.

## 14.3   What design has to do

*Design stages*

Design is carried out at various stages of the project's life-cycle and is usually the predominant activity at the front end. However, the activities that happen at each stage are quite distinct. Although it should be noted that there are several different models of the design stages, these stages may generally be depicted as shown in Figure 14.2, and as described below.

(1)   *Concept*. Various options are generated which will provide the design necessary to solve the engineering problem. The proposals are at a low level of detail, but are sufficiently well developed to be costed using a global estimating approach.

(2)   *Feasibility*. The feasibility of the various solution options are considered with respect to a number of criteria, typically including the cost to deliver the solution, the amount of time that would be needed to deliver the solution, the capability of the implementation team to deliver the solution, whether the solution is congruent with the technology strategies of the project's participants, and what

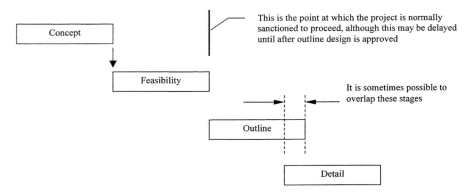

**Figure 14.2** Typical design phases for an engineering project.

environmental impact the solution will have. There are, of course, many more criteria that may be used to assess the design solutions proposed.

(3)   *Outline.* The concept design, that may or may not have been further extended during the feasibility stages, is now developed to the outline level of detail. The major work packages are defined in terms of form, function and delight. This may include process design, space planning, general arrangement drawings, system architecture, etc. A decision will be made to progress the outline design to the detailed design stage.

(4)   *Detail.* The outline design solution that has been approved is now designed in detail. The work packages that make up the overall solution are broken down still further to the greatest level of detail. Individual components are designed, and then integrated to form the work package. The work packages are then integrated to provide the overall solution.

Design does not only occur on projects to build new facilities and infrastructure. Design is also carried out when facilities are extended or refurbished. These design projects can be huge undertakings, as when oil refineries have new process plant capability added to existing facilities, or they may be much smaller in scale, requiring only a few designers to complete the design task, such as extension works on a private house. However, the stages of design are almost always followed as outlined in Figure 14.2. As with design work for new installations or infrastructure, larger refurbishment design projects usually have a project manager assigned to the design group.

### *The product and work breakdown structures*

The processes required to define the work that is required to be done by the designers working on the project will result in a:

❑ product breakdown structure (PBS), which identifies the products, or components, that together form the design solution;
❑ work breakdown structure (WBS), which identifies the work required to deliver the designs needed to produce those products or components, and what resources (human and machine) will be needed to undertake the work.

Some industries also generate an assembly breakdown structure (ABS) that shows how the products or components are assembled together to create the design solution.

The process used to tie the PBS and WBS together (and ABS if used) is the creation of the organisational breakdown structure (OBS). This describes the way that the members of the design team will be organised to carry out the work needed to produce the designs required for the products identified in the PBS. The mechanisms of PBS, WBS and OBS are common to the project management method, and this is unsurprising since creating the design needed for an engineering project is a project in its own right. Design groups frequently have project managers assigned to them in an attempt to ensure that the design work is delivered to the client according to schedule.

## 14.4 The role of design management

Design management is a relatively new area of professional interest. It has only been recognised as an important subject since the early 1970s, and postgraduate qualifications began to be offered in the 1980s. In fact, effective design management within the project management context is still not well understood.

The creative aspect of design work does not easily lend itself to being managed with the same mechanistic focus that can be applied to engineering projects when they move into the implementation stage. As the models of creativity and design discussed in the previous section of this chapter illustrate, the creative element of the designing process requires a period of synthesis that cannot always be 'forced'; the subconscious mind needs to work on the problem. The amount of subconscious activity needed may be little or great. This can appear to make the time management, and hence cost, of design work an impossible task. Indeed, many designers resent the 'imposition' of a mechanistic management regime, since they feel this constrains their ability to design effectively. Despite this, there are a number of design management models which have proved to be effective.

Design management is defined in many ways, from the effective control of the flows of design-related information between the various project participants, through an approach involving control over the design process itself, to a more abstract view relating design to corporate strategy. Since within engineering projects design can be considered to be a project in its own right, then it would seem that project management techniques ought to be applicable to design management. However, managing the classic project-management triple constraints of time, cost and quality is insufficient. Since a fundamental aspect of design is the element of creativity, and the difficulty this brings in terms of an accurate estimate of the time needed to complete a design task, this must be considered in the management regime. Project management is not only about managing time, cost and quality though. It is about understanding the impact of the environment (in the widest sense) on the project. Part of the design project environment is the creative aspect of design work, and hence using a truly holistic project management approach to design management automatically allows for the somewhat unpredictable nature of the design task.

Designers may be classified by the industrial sector in which they are working, e.g. aeronautical, naval architecture, chemical, food, and so on. The sector is then broken down again, with specialisation in each, e.g. airframe, hydraulics, fly-by-wire, or propulsion in the aeronautical sector. Thereafter, the designers will fall into one of several professional disciplines such as mechanical, electrical, control or electronic, and all have their own professional institution to represent them and ensure that a universal standard is applied to the education of these design professionals. There is therefore a likelihood of designers in these disciplines acting in a 'tribal' manner. However, experience shows that this tribal behaviour, if effectively managed, can lead to significant innovations in the design stages of projects, and does not necessarily have a negative impact on the organisation of the design. These culturally driven misunderstandings can lead to a lack of predictability, imaginary and real, that for those attempting to manage the delivery of design can be difficult to deal with.

In addition to the difficulties of lack of empathy between different design disciplines, there is an equally difficult, culturally influenced misperception between designers and those responsible for implementing (in other words building or constructing) those designs. In engineering, as in most other fields of endeavour, the people charged with implementing the design solution have a different mental attitude to that of the designers, making identification between the designer and the manager difficult. That design delivery should be managed must, however, be unquestionable, and this must be either by implementation people or by designers.

The inherent difficulties created by the difference of perception between designers and those charged with building the artefact or facility that the project was initiated to provide is not necessarily overcome by physically integrating designers and makers. The best solution to the problem of the location of these two groups, and the design of their overlapping work processes, is determined according to the circumstances of each project. Projects with nearly identical design and implementation processes, which may even be at the same geographical location, will often have completely different organisational arrangements.

Since engineering designers and those who manufacture, produce and construct are in many cases educated together (and are unlikely to be firmly set on the course of designer or implementers at that time), these differences in perception must be formed, at least in some part, in the workplace that they eventually move in to. One of the important exceptions to this 'co-educated' system is that of architects, and it is likely that the seeds of the antagonism that exists between the architectural profession and the constructors who build their designs are planted during their separate training.

This difference in mental attitudes between the two groups may be visualised as shown in Figure 14.3. The designer works in the area to the

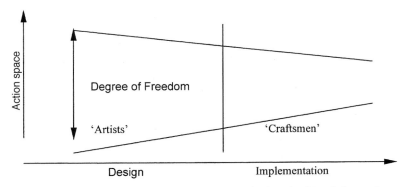

**Figure 14.3** The contraction of the action space during the life of the project.

left of the diagram, where there is a great deal of freedom. This allows multiple perspectives to be taken on a problem, and many alternative solutions to be generated. As the project lifecycle moves from concept design, through feasibility studies, to outline and then detailed design, the degree of freedom gets smaller as the final design solution becomes clear. As the project moves into the implementation phases, the degree of freedom becomes increasingly smaller, and the actions required by the implementation people correspondingly more prescribed.

Therefore, the need for the project participants to understand the cultural differences between design and implementation is important and is not necessarily intuitive. The manager of these design processes must be able to work effectively at the interface between 'artistic' designers and 'craftsmen' implementation people. The person managing design, and seeking to influence the designer's environment within the project, must be able to understand the perceptions of both groups of the project's reality. This position between design and implementation allows the manager of the design organisation to have a significant influence over the amount of design creativity within the confines of the project. This role can be seen to some extent as one of nurturing, or at least supporting, creativity within the design organisation. Sensitivity to the designer's need for autonomy and the management's focus on design (and hence project) objectives is a prerequisite for a successful manager of design.

One of the fundamental requirements for effective management of design is an efficient flow of information between the participants in the project. This applies particularly to those that have an input to the design phases of the project, and a list of such participants would include at least:

- the promoter and appropriate groups within the promoter organisation (such as their design department);
- the users/operators of the project deliverable(s), which may or may not be part of the promoter organisation;
- the project manager;
- team leaders in the project team;
- the design manager;
- the lead designers;
- the design team leaders;
- sub-designers and appropriate team leaders within those groups;
- design approvals consultants acting on behalf of the promoter;
- design checking consultants;
- local authorities;
- statutory bodies.

The length of the list indicates the prospective complexity of the paths (channels) along which information needs to flow, not only in one direction (asymmetric), but also in two directions, since feedback on the content of the information is vital. For instance, for designers to fully understand the brief they have been given requires that they question the brief in order to gain a proper understanding of the brief writer's true meaning. Hence, the briefer and the briefed must have a two-way dialogue (symmetric). This also means that the communication path (or channel) must be duplex, and not simplex, i.e. information can flow in both directions at once (as in a telephone line).

It is important that an explicit communication strategy is developed, and the necessary channels between the project participants are established. Rules for the use of these channels must also be put in place. These rules will cover such issues as the medium in which certain types of information should be transmitted (e.g. paper, verbal, e-mail, video), standard formats within each medium that should be adopted (e.g. how to lay out minutes of meetings), to whom the information should be passed within each channel, and the levels of authority on what information is passed. There are also many other considerations that are dependent on the circumstance of each project. The important thing to remember is that communication must be managed. Always consider that the information being transmitted is important in terms of content, and that the correct person receives it.

The information being discussed in this context is that which facilitates design activity. There are also other types of information that are related to the monitoring and control aspects of the management of design, such as reports, scheduling diagrams (for example, Gantt charts, dependency networks), and document control processes. The control of these types of information is also important for the effective management of the design phases of the project, and forms a significant part of the communication strategy.

## 14.5  Managing the project triple constraints

The management of time in the design phases is fundamental to the delivery of the design at the appropriate time to allow the project schedule to be adhered to. The first part of this chapter has described the inherent difficulties in managing time in the design process. Delivering design information to those needing it, when they need it, is managed using several different processes.

The concept of gates in the project life-cycle is relatively new, and has in the main been learnt in engineering projects from the way in which new products are developed for the consumer market. A gated design process means that at certain points in the design phases, the design must 'pass through' a gate. The rules for passing through the gates must be established, as must the points at which the gates are positioned in the design process.

There are commonly two types of gates. They may be 'hard' or 'soft'. A hard gate is where the design cannot progress to the next stage if the gate is not passed. The design process may not move into the following stage until sufficient reworking has been done to allow the design to pass through the gate. Soft gates are ones in which the design is allowed to progress to the next stage even if it has not been accepted as 'compliant' (dependent on the gate's rules). However, a commitment must be made by the person responsible for the design (the design team leader or the design manager) to make changes to the design to ensure that it becomes compliant before the next gate. It is also possible to have 'fuzzy' gates, which are essentially a combination of hard and soft gates. In a typical fuzzy gate process, parts of the design may progress to the next stage (those that comply with the rules), whilst the non-complying parts must be reworked in the previous stage until they do comply.

Figure 14.4 shows where the gates are commonly positioned, although different organisations, and different projects, will position the gates at different places in the process. Figure 14.5 describes the different types of gates.

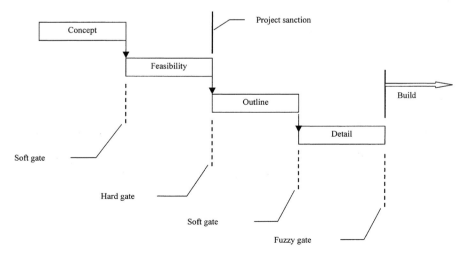

**Figure 14.4** How a gated design process may be set up.

**HARD GATE**

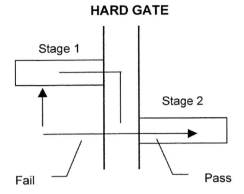

*Rework within previous stage. All design moves forward, or all design is rejected*

**SOFT GATE**

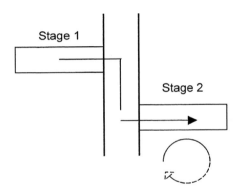

*Rework within next stage, all design moves forward*

**FUZZY GATE**

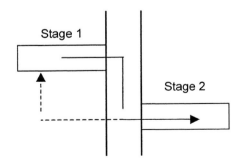

*Rework within previous stage. Complying part of design moves forward, non-complying part reworked in previous stage*

**Figure 14.5** Types of gates.

The process for scheduling design work must be based on knowledge of the individual deliverables (drawings, calculations, reports, specifications, and other documents) from the product and work breakdown structures. Scheduling the creation of the deliverables can then be done from the dependency between the deliverables. Creating a dependency network, and then a Gantt chart from the network, is the most effective way of building a schedule. Smaller design projects can be scheduled by creating a Gantt chart without first creating a separate dependency network. The duration of each individual task in the network is decided after consultation with those with knowledge of the work to be carried out. The difficulty with this type of scheduling tool is that the iteration that is a fundamental aspect of the design process cannot be modelled. The amount of iteration required between design tasks, or groups of design tasks, must be 'built-in' to the schedule in some way. This is not a very satisfactory way of scheduling, and leads to continual adjustments of the schedule as the iterative cycles in the design work unfold. There is one particular method of scheduling that can overcome this difficulty that involves a technique called Dependency Structure Matrix (DSM). However, there is still no commercially available software application that can carry out DSM.

An aspect of planning in the design stages that is crucial when estimating the time and manpower requirements is the degree of front-end loading that will be employed. Front-end loading refers to the practice of employing a significantly higher number of designers earlier in the design phases of the project than has normally been the case. The idea of doing this is to concentrate the project resource at the point where it has the most effect in shortening the overall project time scale. In broad terms, the phase at which the biggest influence can be brought to bear on reducing uncertainty later in the project is at the outline design stage. However, the loading of extra resources is also done at the concept, feasibility and detail stages (Figure 14.6). It must be remembered, though, that at the concept stages highly experienced designers will be employed to identify as quickly as possible the most advantageous options for solving the design problem(s), and there is probably a scarcity of such design resources in any firm.

Putting extra resources into the work at the outline stage allows the development of the chosen design solution to be carried out much more quickly. At this stage, the iterative cycles can be moved through rapidly, and the final outline design solution can be articulated in a much shorter time. This means that the uncertainty that surrounds this stage of design (i.e. will the outline design actually fulfil the promise of the concept chosen at feasibility) is removed from the process much sooner. Detailed

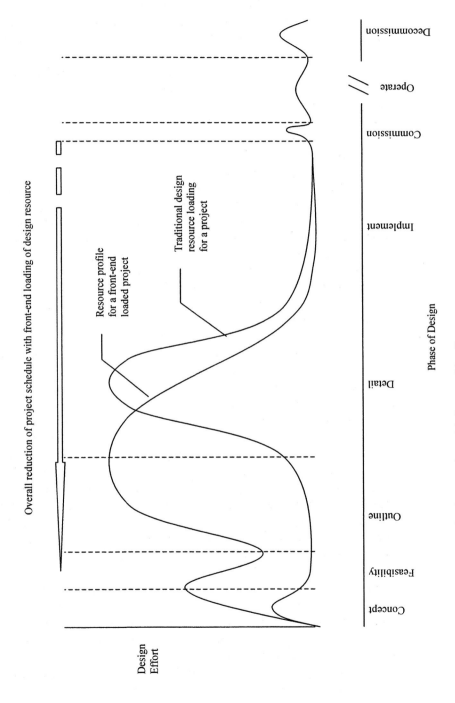

**14.6** The design resource profile for a project with front-end loading.

design can then be started with less risk that the outline design will have to be revisited (which often means that the design process has to be stopped while the implications of a technical risk in the outline design are reassessed).

The cost to deliver design work is almost entirely the cost of the time of the designers assigned to produce the design – essentially the cost of the human input to the process. There are overheads to consider, such as the cost of the facilities required to carry out the design (the design office itself, power, design tools such as CAD, software packages), administrative support and management overheads, but the predominant cost is the designer man-hours required to produce the deliverables. This cost can be estimated using the information contained in the schedule. The time required to complete a design task can be multiplied by the hourly rate for the appropriate skill level of design required, producing the cost to complete that task. Overheads can be added per task, or added to the overall estimate once all the tasks have been costed.

Once the cost estimate to carry out the design work has been completed, the next stage is to consider the risks associated with the design tasks. These may include such issues as: How well known is the technology that is being designed? If the technology is mature, and there is great experience and knowledge in the design firm of working with that technology, there is probably little risk in this area. However, if the technology is new, or the designers have little experience of working with this technology, then the risk of overrunning the time allowed to produce the design deliverables is high. Other risks may be associated with the lack of stability of the design brief (or user requirements if these were issued instead of, or in addition to, a brief), and the likelihood of changes to the brief from the promoter. There may be unknown risks from the construction process, classically, in the construction industry, that of the actual ground conditions that will be found as opposed to the conditions predicted from the geophysical investigation. There are a large number of risks that could have an impact on the design schedule, and hence the cost estimate. These must be assessed, and the appropriate contingency must be made in the schedule and cost estimate. It is most helpful if the design organisation, meaning all the firms with a significant input to the design process, are involved in the overall project risk management process. Risks to other elements of the project are often not identified as having a possible effect on the design process. Design involvement in the risk management process can help to ensure that these risks are picked up and incorporated into the design schedule and cost estimate.

Once the estimate has been produced, and accepted by the management function of the design firm(s), it is turned into a budget. This is an allocation of money to the manager of design to pay for the design work. At this time, the firm becomes committed to the expenditure, and the manager of design now has to provide accurate cost information to the accounting function to allow effective control of the business to be carried out. Cost control on larger design projects may be done using earned value techniques, and this methodology is discussed in Chapter 10. Often, though, design cost control is based on simple, and not integrated, cost reporting against the scheduled progress of work.

It is during the design phases of the project that much of the quality of the ultimate project is established or enabled. There are a number of definitions of quality, but most of them in some way or another address the need to meet the requirements of the promoter.

Two clearly different quality processes must be encompassed in the design phases of a project. The first is to ensure high-quality design work, in and of itself. Doing so is not a straightforward process, but consists of a number of aspects. Accurately *capturing* the requirements of the promoter is fundamental. Putting in place a design *process* capable of developing an appropriate solution is essential. Ensuring that the solution developed *satisfies* the promoter's requirements is critical. Carrying out these activities effectively is dependent on approaching the tasks with a consistently good attitude towards achieving high-quality design. The second process is about enabling the construction, building or manufacturing (the implementation) of the design to be carried out to high-quality standards. This entails designing within the process capability of the implementation phases of the project. Process capability is well established in manufacturing industries, where the production machinery's capability to make a component to the required accuracy must be considered in the design stage. This means, for instance, that designing a component to a machining tolerance which is not achievable (by the equipment used to make the component) is explicitly prevented by the design processes used. This concept of process capability is directly analogous to ensuring that the design for a structure is buildable *in practice*, as well as in theory.

The most well-established and comprehensive quality system for the design phases of a project is known as quality function deployment (QFD), or the 'house of quality'. QFD can be considered to be a system for designing. It monitors the transformation of the promoter's requirements into the design solution, to ensure that quality is inherent in the solution. To do this, QFD integrates the work of people in the project's participant organisations in the following areas:

❑ briefing (to understand what the promoter's requirements are);
❑ engineering (to understand what technology is available);
❑ implementation (e.g. manufacturing, construction – to understand the process capability);
❑ marketing (to understand the promoter's perceptions of the solution that satisfies the requirements);
❑ management (to understand how the processes to ensure quality can be put in place).

The primary set of considerations for the QFD team are given below.

(1)    *Who* are the promoter, in the broadest terms; i.e. the users of the project deliverable, the owner, other stakeholders.
(2)    *What* are the customer's requirements, which may or may not be explicitly stated in the design brief.
(3)    *How* will these requirements be satisfied, including an evaluation at the highest level of abstraction, such as should the project actually build a road or a railway to meet the requirement to transport people from A to B?

The term 'house of quality' comes from the shape of the matrix used to capture the information generated in the QFD process (Figure 14.7). The client's requirements are scored in order of their relative importance and ranked after a weighting criteria has been used. The 'roof' of the matrix contains the elements of the design solution that will satisfy the

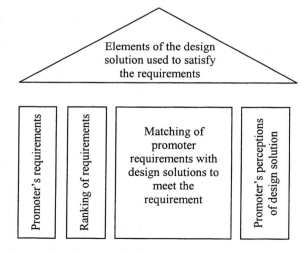

**Figure 14.7** The quality function deployment matrix.

requirements of the client. The ability of these elements of the design solution to satisfy these requirements is then estimated using experience and judgement in the central core of the 'house'. The final part of the QFD exercise is to evaluate the client's acceptance of the design solution in total, and at the elemental level of the matrix, with regard to their requirements. It is important to understand that the QFD will not generate a design solution, but it will allow the quality of that solution in meeting the promoter's requirements to be monitored with rigour and accuracy.

## 14.6  Design liability

A key consideration in the management of design is the issue of the legal liability of the designers for defects in the solutions they create. Failure of the design solution may have many repercussions and affect many people. The designer may be liable under many different areas of the law. These fall under the three main groups of contract law, tort and statute law. There is insufficient space in this chapter to make any attempt at a meaningful coverage of the issue of design liability, so only a few commonly found issues will be touched on.

Designers providing a design solution to a promoter with whom they have a contract are required to comply with the terms of that contract: usually to provide a design within a certain time, to an agreed price. In addition, the contract will specify technical details of the designer's work, against which it will be possible to determine whether the design solution fulfils the contract between the promoter and the designer. Failure to meet the terms of the contract is known as a 'breach of contract', and is pursued in the civil courts to enable the promoter to get recompense, usually in monetary terms, for the damage inflicted by the failure of the design. The case will hinge on the terms of the contract (unless overridden by statute law, for instance where health and safety have been compromised).

The designer may also be held liable in tort, which is a non-contractual liability to third parties. When designers produce a design solution that is faulty they may, in legal terms, be negligent, but only if they did not take reasonable care when carrying out the work. The definition of negligence is that the designer must use 'reasonable skill and care' in the execution of the work. Common law (under which most contract and negligence law falls) is based on judging the present case against the principles laid down in previous similar cases in which a 'precedent' has been set by a judge's ruling. There are appropriate precedent cases in most sectors of industry that relate to the negligence of designers. There are also two particular cases that relate to the term 'reasonable skill and care' of a professional (which is what a designer is considered to be). The first is known as 'Bolam's test', after a case in which the reasonable skill and care of a doctor, called Bolam, was tested in a negligence case. Bolam's test sets the precedent for the skill to be expected from an average person working in their field of professionalism. The second case is called 'Roe v. The Ministry of Health' (1954). This case established that a professional cannot be liable for negligence if they are following the normal practice at the time, even if that practice can subsequently be shown to be flawed. This has importance for designers working on innovative design solutions which have not been tried before. In these situations, despite taking reasonable skill and care, a design fault may occur that can only come to light after failure of the design; it could not have been predicted in advance. In this situation, the designer is unlikely to be found negligent.

It is usual for designers to have professional indemnity insurance in place, in case they are found to be have been negligent in the execution of their work for a promoter with whom they have a contractual relationship. However, this type of insurance is expensive and does not buy a great deal of financial protection. For this reason, contracts for the supply of design usually have strictly limited liability clauses, with the designer (the supplier) often accepting liability only up to the value of the contract.

Another very important issue for designers is their responsibility for 'product' defects. In the area of product liability, which is most commonly found in products for the consumer market, the law does not allow for the professional designer to exercise reasonable skill and care. Liability in this case is 'strict'. This means that the designer is liable for any defect in the product that is there as a result of the design process. Professionalism, or the lack of it, is not an issue. If there is a defect in the design, the designer is liable. In reality, the designer's exposure to risk due to a design fault is often submerged beneath the other product liability issues of manufacturing, workmanship and materials liability.

Health and safety law is also an important consideration for designers. There are general industry-wide requirements for designers to produce safe design solutions, particularly (but by no means limited to) where the general public are exposed to the dangers of inadequate design. Many industrial sectors also have specific legislation to protect workers and the public from design failure (an example of which is the Construction (Design and Management) Regulations 1994).

The importance of assessing and then controlling the design liability issues by managers of design is clearly of great importance. The design manager must be aware of the safety, contractual and insurance implications of the design project to ensure that they are managed effectively.

## 14.7 Briefing

The brief is often considered to be the key document for a design project. Although most sectors of industry would recognise the term, some do not: software design houses are more used to working with a statement of requirements, although the basic concept of the document is similar. Stated simply, the brief is the method by which the promoter (external to, or within the same organisation as, the designer) tells the designer what it is that needs to be designed. This sounds pretty straightforward until one begins to articulate exactly what is wanted by the people commissioning the design. The briefing process can, in fact, be excruciatingly difficult. The main cause of the difficulty is the way in which a solution to a problem is arrived at. The iterative and uncertain processes involved in arriving at a design solution were discussed earlier in this chapter. It may be considered that the ideal place from which a designer ought to begin working from is a pure statement of the requirements of the promoter (the requirements of a promoter are a clear and concise statement of the problem which the design is to overcome, completely devoid of any suggestion of the solution). Unfortunately, things are not so simple.

First, it is enormously difficult to state requirements without some reference to the possible solutions. The requirements-capture process has only become better understood in recent years, and reliable models for requirements capture are not always easy to find. Some are available, however, although they tend to be associated with specific industries such as automobiles or information systems. To carry out the process properly is often a costly and time-consuming exercise, as the information required to generate a complete picture of the requirements must come from many people (users, builders, suppliers, marketing, maintenance engineers, and other stakeholders). The time and money to do the work may not be available.

Second, just as how to create 'solutionless' requirements is becoming better understood, some authoritative design experts are cautioning against the creation of such information at all. It is frequently found that the briefing process is distorted when the initial design concepts are presented to the promoter, and the promoter says, 'I don't want that!' The initial brief is then adjusted by the promoter (hopefully with input from the designer) to reflect what type of solutions are not acceptable to the promoter. Sometimes this cycle can extend far beyond the original design period, costing the promoter much more money than originally envisaged to employ the designer, and create great tension (and probably reduced motivation) in the designer or design group.

Third, to make the briefing process even more opaque, it is also being acknowledged that one of the cardinal rules of the briefing process can legitimately be broken. The rule of never changing any of the requirements is often considered to be inviolate, and fundamental to the existence of the brief. This is frequently being challenged in some sectors, particularly in new product development, where the market for a particular product can be as short as a few months. Changing, or relaxing, a requirement in a design brief in these situations can mean that a design solution is created that will get the product in the market earlier than competing products, and may mean the difference between the organisation surviving or failing.

The simple fact is that briefing is less about a repeatable process, although this may be desirable in many design project situations, and more about the interactions between the brief writer, those responsible for determining the requirements upon which the brief is based, and the designer. Creating a brief from which the designer can work most effectively, and in which the promoter has the most confidence of an appropriate design solution resulting, is a joint exercise. Frequently, this does not happen. Many promoters believe that they alone are capable of creating the brief, even if they employ specialist design expertise to help them do so. The point is that the actual designer who is to provide the design solution should have a big input to the brief, because the way he or she works will have an impact on the types of solution that will be created, and this can then be reflected in the brief. Involvement in the brief writing will also enable the designer to understand far more completely what it is that the client is likely to accept as a solution. The three parties together are also much better equipped to challenge the requirements definition feeding the brief, and seek sensible changes in the requirements based on information they all bring to the process. This means that less iteration between the promoter and designer is likely, although the design process iterations will still be required as part of the natural evolution of the final solution to the problem. The design outputs stand a far better chance of being delivered within the time scales envisaged by the promoter and at the cost expected.

One of the many check lists for creating a design brief is included here for reference. It was designed by the Department of Trade and Industry and published in 1989. It provides a useful basis from which to assess the needs of a brief in any industry.

❑ Value for money/lifetime operating costs
❑ Product uniqueness/superiority
❑ Selling price
❑ Performance
❑ Reliability
❑ Serviceability
❑ Maintenance costs
❑ Life expectancy
❑ Versatility
❑ Running costs
❑ Ease of operations/user appeal
❑ Ergonomics
❑ User friendliness
❑ Appearance
❑ Legislative and community factors
❑ Compliance with regulations and standards
❑ Safety
❑ Environmental impact
❑ Factors important to the manufacturer
❑ Time-scales of the development programme
❑ Unit cost of production
❑ Facility for future range expansion or product improvement
❑ Balance between in-house manufacture and bought-in items
❑ Other factors
❑ Size and weight
❑ Ease of transport and installation
❑ Cost of the development programme
❑ External consultancy requirements and need for the involvement of sub-contractors or suppliers
❑ Production levels envisaged and organisational implications
❑ Investment requirements – for stock, work in progress and production capacity.

(DTI, 1989)

## 14.8  Interface control

A design solution very rarely has a single discrete product, or is comprised of a single component. Far more usually, the solution comprises a significant number of components, and most often there are a significant number of components arranged in a number of sub-systems. For instance, a software programme of any size at all will be made up of a number of modules of code, that act together to create the program's functionality. A car engine is comprised of many sub-systems: the fuel supply, the oil supply, the cooling water, the electrical system, and so on. This property of almost any system is not limited to engineering. Sophisticated drugs similarly comprise a number of systems of inter-acting components, such as the material used to bind the 'active' parts of the drug together, the complex series of components that form the active component, and often a special outer coating which itself is a complex formulation. In almost all situations where designers work they will be dealing with a design solution that comprises multiple sub-systems, each containing multiple components. The management of the interfaces between these sub-systems is often crucial to the effective creation of the solution. Some disciplines manage these interfaces better than others, and those that do it well and consistently are usually 'system' orientated, e.g. electronics, information systems and aerospace. Other sectors such as heavy engineering and civil and structural engineering, are far less systems focused despite the self-evidence of the fact that they also create systems (a bridge is clearly a system of interacting sub-systems).

Effectively managing the interfaces between the sub-systems creates significant advantages in the overall management of the design process. Setting, and subsequently 'freezing', the interface requirements between sub-systems means that the designers of the sub-systems can then work on designing their part of the overall solution without further reference to those working on adjoining systems. Each sub-system design only needs to satisfy the interface constraints. If these are met by the sub-system, then the operation of the internal components in the sub-system is not relevant to other interfacing sub-systems (from which property the term 'black box' arises). Hence, the need for information to flow con-stantly between designers working on separate systems is removed.

The work of defining interfaces is not trivial. In a system-orientated design solution the 'architecture' of the overall solution (the way in which the sub-systems 'fit' together) is usually the responsibility of a senior, experienced engineer. The architecture determines the way in which the design solution is broken down into manageable sub-systems, and what, and at what level, the interface constraints are set. The degree to which the overall design is broken down, and the size of the sub-systems, is fundamental to an effective system, and hence design, man-agement. The crucial interface management issues are given below.

(1)   Deciding to what level the overall design should be broken down into sub-systems, and therefore determining the number and 'positioning' of the interfaces.

(2)   Ensuring that the setting of the interface constraints reflects the needs of the overall design solution. This means that compromises will need to be made for individual interface constraints.

(3)   Deciding on the tolerances that the constraints should have. If the constraints are too tightly specified, optimisation of the sub-system design can be reduced dramatically; if too loosely specified, the overall design solution will probably perform poorly.

(4)   Ensuring that the interfaces, and their constraints, are 'frozen' at an appropriate time in the design project's life-cycle. Freezing too soon will lead to sub-optimisation of the overall system, since not enough is known about the system's properties, whereas freezing too late will prevent the designers from making the technical (and quite likely commercial) decisions needed to deliver the sub-system on time.

A schedule of the interfaces showing freeze dates and required delivery dates for sub-system designs is a valuable management tool. It will also be needed for configuration management – broadly, the process of ensuring that the system architecture is allowed to change in a strictly controlled manner.

## 14.9  Design for manufacturing

In most sectors of engineering, most of the cost in the project's life-cycle is incurred in the implementation stages: in construction when concrete is cast into its form in the position where it will remain; in mechanical and electrical engineering when the designed components are manufactured; in electronics when the circuitry is assembled. The 'making' stage of a project typically accounts for between 75% and 90% of the total cost of the final deliverable project. Therefore, anything that reduces the cost of manufacturing, or producing, the components is worth pursuing. One of the biggest cost drivers in manufacturing is design that does not take account of the most cost-effective processes for making the components.

It is clear that it is vital that manufacturing specialists have a significant input at the design stages. The process of bringing in this expertise to design is called design for manufacturing (DFM). In essence, the design for the processes that will be used to make the components is optimised at the earliest stages. This is not an easy or comfortable approach to design for either designers or manufacturing specialists. Reference to Figure 14.3 indicates the fundamental difference in mental models between designer and implementer. Getting these groups of people to work together effectively is a key task of the design manager when the design for manufacturing is being done. It is important to recognise that for the greatest effect, DFM needs to be started at the earliest stages of design, when concepts are being generated for the various solutions to the design problem. There is little point in choosing a concept design to progress into detailed design work if the concept chosen cannot be supported by the existing capability of the organisation to make the components. At the lowest level, DFM allows a logical debate to take place about trading-off the costs of new manufacturing capability against the attributes of the design that can create extra value in the final product.

The success of DFM can be ensured by recognising, and acting on, the realisation that differing cultures exist within design and manufacturing. The primary obstacle that this difference creates is that of effective communication. There are two key ways to improve communication between these two groups. The first is to plan for it to happen. This means identifying where in the life-cycle of the design project DFM will have the most effect (invariably early on), and then ensuring that appropriate DFM processes are created in time to be used most effectively. It also implies that DFM workshops and review meetings are built into the schedule. The second way to ensure that communication happens is to create some common understanding of the issues faced by the two groups. For instance, it is far from obvious to designers that the manufacturing process capability required to make their design solution may not exist, particularly when an external promoter is doing the making. However, this lack of knowledge about manufacturing capability is also frequently found when the design will be made in-house. Equally, manufacturing specialists are rarely aware of the specific reasons why a particular feature of the design is necessary to create added value to the promoter. These differences in awareness between design and manufacturing are natural and to be expected. It is up to the managers of design to manage the DFM process proactively for the greater good of the client and, ultimately, the design organisation itself.

A related design management process is design for assembly (DFA). A major part of the 'making' cost for a design solution is the time needed to assemble the various components forming the overall product. In such industries as aerospace, power engineering, consumer electronics, and the manufacture of white and brown goods, the assembly time is heavily influenced by the ease of assembly of the product that will be sold. Consequently, the specialists in assembly processes must be brought into the design process in the same way as the manufacturing experts are involved in DFM. Unsurprisingly, the differences in culture between the designers and the assembly specialists is just as evident in the DFA process as for DFM. Communication between the two groups is facilitated in the same way as for DFM, i.e. plan to communicate, and create a situation where a common understanding can be reached.

The management processes of DFA and DFM also interact with the design solution. It is quite possible that the design of a component that has been optimised for manufacturing is very difficult to assemble, adding time (and therefore expense) to the processes that will deliver the final product. Conversely, a design optimised for assembly may be expensive, or even impossible, to make using existing manufacturing process capability. It is incumbent on the design manager to ensure that the correct trade-offs are made between designing for maximum client value, low-cost manufacturing, and ease of assembly.

## Reference

DTI (1989) *Design to Win: A Chief Executive's Handbook*, Department of Trade and Industry.

## Further reading

Allinson, K. (1997) *Getting There by Design: An Architect's Guide to Design and Project Management*, Architectural Press, Oxford.

Cooper, R. (1995) *The Design Agenda: A Guide to Successful Design Management*, Wiley, Chichester.

Eisner, H. (1997) *Essentials of Project and Systems Engineering Management*, Wiley, New York.

Hamilton, A. (2001) *Managing Projects for Success*, Thomas Telford, London.

Lawson, B. (1997) *How Designers Think*, Architectural Press, UK.

Reinertsen, D.G. (1997) *Managing the Design Factory*, The Free Press, New York.

# Chapter 15
# Supply Chain Management

This chapter defines supply chain management, drawn on theory and research studies across a range of industries to explore the implications for the project manager. The chapter presents a model for supply chain management.

## 15.1 Background

Supply chain management is a strategic function of the firm that integrates those external and internal activities required to manage the sourcing, acquisition and logistics of resources essential for the organisation to produce products or services that add value to its customers.

There is considerable ambiguity about the terms and definitions and the scope of supply chain management. This chapter presents the terminology used in the field, and looks at the concept of supply chain strategy, including highlighting the influences on the purchasing and supply functions. The notion of the 'world-class' organisation is introduced, and the implications for construction are drawn from this. The chapter introduces the idea of a supply chain system comprising a project-focused demand chain generated by the promoter and the contractor, and acting as a multiple project-focused demand chain hub that has to develop a supply chain strategy to meet different needs. Within this idea, the implications of different types of promoter are explored, and the concept of the project value chain introduced. A recently introduced procurement route initiated by the Ministry of Defence (MoD), and termed prime contracting, is discussed. The chapter goes on to explore research into supply chain management that builds on work related to the MoD's prime contracting initiative. The chapter also presents research work conducted on supply chain scenarios in the constructional steelwork sector as an example of the potential impact of the newer procurement routes such as prime contracting and the private finance initiative (PFI). It suggests that a new role will emerge in the industry, that of the strategic supply chain broker, who will compete on the basis of core competencies in supply chain management. It also explores the possible restructuring within the sector that may occur due to the emergence of the broker role. The chapter concludes with a section that develops a framework for the supply chain system in construction.

## 15.2 Perspectives on terminology

The 'traditional' model of purchasing and supply focuses on developing and retaining appropriate knowledge and skills in the purchasing area. Hence, typical elements in the traditional model include a specialist department, or section within a department, dealing only with purchasing and the placement of orders with suppliers. It operates within a known hierarchical organisational structure, where paperwork systems dominated task activities prior to the advent of computers. Policies and procedures will have been established to deal with the enquiry and competitive bidding processes, order placing and contract management. 'Price' will figure strongly when purchasing managers evaluate a supplier's performance, and decision making will focus on securing the right quality, quantity, price, supplier and location, and delivery at the appropriate time. The focus will also be on the individual transaction, although some repeat purchasing may occur. Finally, relationships with suppliers will predominantly be of an adversarial nature and at arm's length, using competitive mechanisms for supplier choice. The traditional model is one where purchasing acts as the interface between the firm and its suppliers under conditions of market-based competition and economic rationality.

Due to competitive pressures, however, a number of recent and significant trends have resulted in a need for organisations to become more effective and efficient, with a consequent influence on the way they are managed, including the purchasing function. Organisations, especially the larger ones, have needed to become more adaptable, responsive and flexible to changes in the business environment. Organisational layers have been removed, with an increased use of cross-functional teams and a delegation of responsibility to lower levels in the hierarchy. In addition, communication is now horizontal as well as vertical, with mangers changing from a directing to a facilitating role. Incentives and staffing systems are being aligned to accommodate these changes. Managers have changed the manner in which they approach the wider organisational context. This is also linked to a greater appreciation of the opportunities that can accrue from more cooperative ways of working between suppliers and customers, resulting in cultural changes within firms. The consequence of this is that firms are more willing to consider working closely with suppliers and customers to create a more integrated production and supply process that goes beyond legally defined organisational boundaries. This means that firms now have to work out their role and position within a wider configuration of organisations, with a consequent impact on organisational structure. This has put the supply function clearly on the strategic, and not the operational, agendas of firms, and potentially opens up new roles, including associated impacts on the marketing function, with 'relationship marketing' coming to the fore.

This different way of thinking about supply and purchasing has resulted in a series of new terms being adopted and used to describe the domain. The *supply chain* concept has emerged, and has an underlying implication of a sequence of interdependent activities that are internal and external to the firm as a legal entity. This can encompass single-location activities for fairly simple firms, to multi-site, geographically dispersed locations, often located internationally.

The idea of the 'chain' has been extended to include analogies with rivers, and 'upstream' and 'downstream' terminology has become infused into the language, often including the concept of a 'supply pipeline'. Further extensions to the concept have included viewing a supply chain in network terms, seeing it as a series of connected, mutually interdependent organisations cooperatively working together to transform, control, manage and improve material and information flows from suppliers to customers and end-users. One implication of the supplier network is that the boundaries of different supply chains might overlap, and particular suppliers might become nodes within a more complex web of patterns of suppliers. Typically, flowchart mapping techniques are used to understand these 'flows' within the chain or supply pipeline. The impact of lead times and individual cycle times is also encompassed within the analysis.

For the supply chain to work as an integrated system requires the management of both materials and information. It also involves managing the upstream and downstream business relationships between customers and suppliers to deliver superior customer value at less cost to the supply chain as a whole.

## 15.3  Supply chain strategy

Product strategies provide the basis upon which a supply chain strategy is built. Product strategies requiring different approaches to time, cost, quality and product innovation place different requirements on the supply chain and its structure and infrastructure. The structural features of the supply chain, as indicated in Figure 15.1, involve the fundamental physical activities of *make – the production activity*, and *move – the logistics activities* and *store*. The infrastructure of the supply chain consists of those features which are concerned with controlling the operation of the physical system, and include planning and control systems, human resource policies and communication strategies.

Design and technology strategies are concerned with decisions about which activities are carried out internally and which are carried out by external suppliers. Design strategies also include *product* and *process* design. Decisions on design and make will encompass those to be undertaken either within the firm or by suppliers. Equally, they will entail related decisions on which design and technology competencies are to be retained internally, and the extent to which supplier innovation would also be encouraged. Design strategy considerations could also include:

- the use of concurrent engineering principles;
- supplier involvement in design teams;
- product simplification and standardisation;
- the use of computer-aided design through the supply chain and the coordination of design processes and data when in use.

The extent of 'Make' or 'Buy' is one of the fundamental strategic questions that managers in a firm have to ask in supply chain management, i.e. the scope of the activities undertaken within the firm versus those that are carried out by external firms. Finalising the make versus buy decision defines that part of the supply chain that is within the firm's direct control, as opposed to that requiring the investment of resources to develop, enhance or retain internal capabilities. Sub-contracting, in its various guises, is an extension of the make-or-buy decision.

Finally, the creation of sub-assemblies within an overall finished product, as part of a sourcing strategy, creates a tiered structure to supply chains. The final 'manufacturer' or 'producer' has a choice. They can decide to purchase and assemble all components and items and then sell on the completed product, or they may decide to break the final product down into major sub-assemblies and contract these out for manufacture. The final producer thus assembles a series of sub-assemblies. Tier 1 suppliers are responsible for producing major sub-assemblies. Tier 1 suppliers, in turn, will source from their own suppliers – Tier 2 suppliers. It is possible that Tier 2 suppliers will have components req

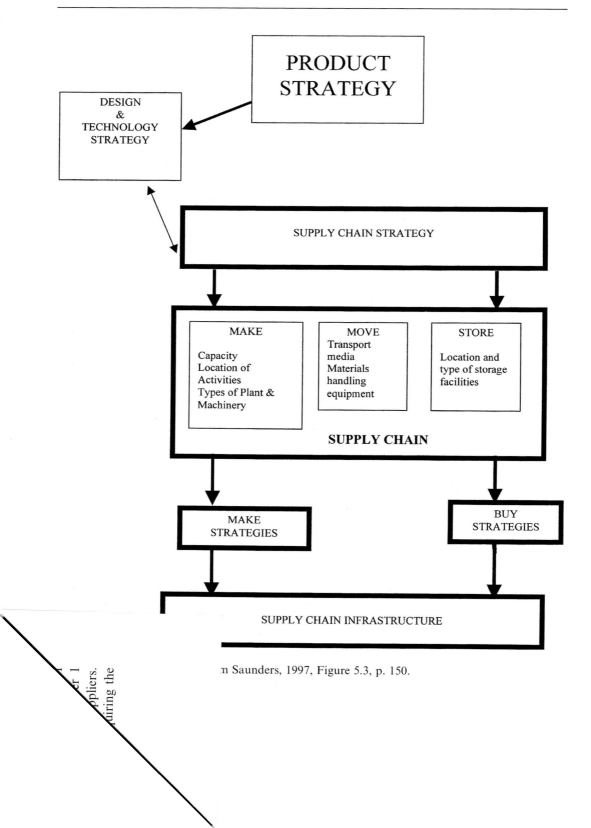

SUPPLY CHAIN INFRASTRUCTURE

m Saunders, 1997, Figure 5.3, p. 150.

inputs of Tier 3 suppliers, on so on. The number of tiers within a supply chain creates its structure and shape.

## 15.4 The nature of the organisation

Saunders (1997) proposes four types of organisation described below.

❏ Those that are in the primary sector, i.e. they are involved in extracting a product or material that exists in nature.
❏ Manufacturing organisations that process, shape, form, transform, convert, fabricate or assemble materials, parts and sub-assemblies into finished or intermediate products for sale to external customers. Saunders acknowledges that within a heterogeneous group of firms there are sub-types that have an impact on the corporate, business and supporting manufacturing, purchasing and supply strategies. The on-site construction process is the equivalent of this type.
❏ Those that are in the tertiary sector and are wholesalers and retailers. This type of organisation has no internal transformation process; it buys in from external suppliers and sells on to external customers.
❏ Those that are in the tertiary sector and are service providers. Here, the product received by the customer is intangible. Design teams in construction are part of this sector.

Saunders has also explored SCM issues through the type of manufacturing organisation. Hill, quoted in Saunders, suggests that there are five process types of manufacturing organisation:

(1)  project based;
(2)  jobbing unit or one-off;
(3)  batch;
(4)  line;
(5)  continuous processing.

This schema accounts for variations in volume, continuity and variety of products being manufactured; each will have a different impact on SCM requirements. Construction is clearly project-based. In addition, factory layout has an impact. Saunders proposes four distinct approaches to factory layout.

(1)  Fixed position, where all resources, workers, machinery and equipment, and materials are brought to a fixed location where the product is built. This typifies construction.

(2)   Functional or process layout, where the factory layout is divided into separate areas or workflow-areas specialising in one type of process.

(3)   Line or flow processes, where materials proceed in a fixed sequence through a series of processes, often dedicated to a particular product.

(4)   Cellular manufacturing or group technology, where separate work centres are established with self-sufficient workers and equipment to make specialised families of parts or products.

## 15.5   World-class organisation in manufacturing

Owing to global competitive pressures, the manufacturing sector has concerned itself with developing the 'world class organisation', i.e. one that has an international reputation for overall effectiveness and knows its core business well. This class of organisation will encompass four critical elements.

(1)   A clear customer focus, normally through using flat organisational structures.

(2)   The flexibility to provide rapid responses to customers and competitors.

(3)   A programme of continuous improvement, which is the key to differentiating the firm from its competitors, achieving sustainable competitive advantage and promoting internal organisational learning. A programme of continuous improvement will normally involve the elements described below.

❑   Benchmarking against competitors, in terms of either products, functions or processes, and using best practice as an extension of process benchmarking or strategic benchmarking in terms of the future direction of competitor firms. Benchmarking priorities in the supply chain revolve around an assessment of which processes within the chain are strategically important, have a high relative impact on the business and have a crucial impact on competitive advantage.

❑   Corporate strategies that are designed to expand the organisation's knowledge assets.

❑   Empowerment of employees and an incentive-led innovation policy.

❑   The use of outsourcing where it adds value.

(4)   The effective use of information technology to provide accurate and

reliable information, to create the ability to respond rapidly and differentiate the firm, and to provide a competitive advantage, greater knowledge and hence an understanding of the market place.

Depending on the procurement route adopted (a project-focused demand chain issue), Figure 15.2 shows that the promoter and the design team generate the major cost commitment for most projects, but are only responsible for approximately 15% of the promoter's expenditure, primarily through design team fees. The contractor is responsible for the major elements of expenditure, usually some 85% of the project cost, and yet, depending on the procurement route adopted, may be cut off from any direct influence on promoter and design-team thinking and their commitment of cost in the early stages of projects. Hence, when looking at a project from a value-for-money perspective for the pro-

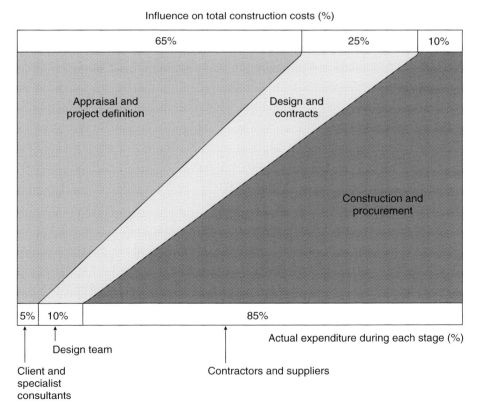

**Figure 15.2** A schematic diagram comparing the costs and expenditure during phases of a construction project. Source: Standing, 1999, Figure 4.8.

moter, certain procurement routes preclude the contractor's knowledge and expertise from being accessed for the benefit of the project, and from potentially adding value much earlier in the project process.

In terms of the roles and responsibilities within the distinct chains, the primary function and nature of the organisation will also have an obvious impact. Design-team firms are clearly tertiary-level service providers whose role can change depending on the procurement process chosen. The main contractors are also service providers offering to manufacture end-products that are designed, purchased and made to order across numerous project-focused demand chains. However, the multi-project supply chain of the contractor has to handle not only this, but also a number of other critical influences, for example multiple fixed-position 'factory' locations that require all resources to be brought together for the assembly process at each location. Equally, where contractors act as manufacturers, they will face considerable product diversity owing to different promoter types, their individual requirements and design-team influences, coupled with the impact of the choice of procurement route on roles and responsibilities. In terms of the capacity and capability to influence their demand chains, regular, knowledgeable, volume-procuring promoters are in a position of considerable power when influencing their own project-focused chain. Some contractors that have the advantage of size will be able to influence their own multi-project supply chains.

### Promoter influences

One important consideration for supply chain management in engineering is the impact that the promoter (or customer) has on the process. Each promoter has distinct requirements and value systems, driven by their own organisational configurations, business and/or social needs for a project, the external environment to which they have to respond, and the manner in which they approach and interface with the construction industry. The promoter commences the process of procurement, brings a demand chain together through a project process, and requires a completed product – a facility of some type – to meet a need.

A number of distinguishing characteristics can be applied to promoters. They can be separated into *public* or *private*-sector promoters. The public/private divide puts different pressures on the project-specific demand chain. Public-sector promoters are now driven by public accountability and best value. Private-sector promoters are much more homogeneous, with influences ranging from the impact of share holder values on projects to time-to-market considerations and ownership considerations due to private limited company or family business status. Promoters also differ in their level of knowledge of the industry, and can be characterised as two types.

❑ *Knowledgeable*. These promoters will generally have a structured approach to project delivery, often encapsulated in some form of project delivery manual or set of procedures or guidelines, including the manner in which they will have developed the project brief for the industry. They will treat the supply chain and its members as 'technicians' who must deliver a project or projects to meet their business or social needs. They will employ either internal or external project managers to act on their behalf as their interface with the industry, and they will tend to be innovative in the manner in which they approach procurement. Knowledgeable promoters will generally be the volume procurers of services, and be demanding of the supply chain.
❑ *Less knowledgeable*. These promoters will often have limited or minimal in-house expertise and knowledge of the operations of the industry. They will rely on the design team to brief the project, and will often approach the consultants first, depending on the type of project. This type of promoter will have limited, if any, appreciation of the complexities of engineering, and will tend to be directed onto a traditional procurement path because of their initial point of contact with the industry.

Promoters or customers to engineering projects can also be classified by type.

❑ *Large owner/occupiers*, who will use facilities as part of their on-going corporate strategic plan to meet a business or social need, and will normally have undertaken an intensive study of their project needs.
❑ *Small owner/occupiers*, who will often react to change, and will be driven to approach the industry because their existing facilities are inadequate in some way.
❑ *Developers*, who view facilities as a method of making profit, will trade the asset to achieve this, or see it as an investment to generate profit and look for business opportunities and available sites to ensure a quantifiable return.

A fourth dimension to promoter characteristics is the economic demand placed on the industry in terms of volume (frequency and regularity), and the extent to which standardisation may exist from project to project in terms of parts, processes and design.

❏ *Unique* projects have a distinctiveness of technical content or level of innovation, or are leading-edge projects that push the barriers of the industry's skills and knowledge to the limit. With this type of project there is limited, if any scope, for efficiencies of process or standardisation and repetition. Typical, SCM tools and techniques suggested by Croner (1999) include:
  ■ the use of competitive tendering coupled with strong pre-qualification and post-tender negotiation processes;
  ■ control over product delivery, exercised through specifications and forms of contract and quality assurance processes;
  ■ a reliance on good professional advice.

❏ *Process* projects can occur where the promoter has repeat demands for a project, and a high degree of standardisation is possible through the volume placed into the industry. Efficiencies can occur from standardisation of design, components and processes. There are many similarities to manufacturing sector assembly lines. MacDonald's, the fast food restaurant promoter, are an example of this type of project. Typical SCM tools and techniques proposed by Croner (1999) include:
  ■ the use of forward planning and demand forecasting techniques;
  ■ rationalisation and consolidation of suppliers by spend;
  ■ the use of strategic alliances, joint ventures and partnering with suppliers using non-contractual forms of agreement;
  ■ the use of performance management and continuous improvement, quality circles, total quality management, just-in-time and inventory management, and lean supply systems.

❏ *Portfolio* projects, where the promoters have large and ongoing spends across a range of project types. However, unlike the process approach, this type will involve a diverse range of needs in terms of technical requirements, degree of uniqueness, or customisation and content, but regular spends will permit the development of long-term relationships with some suppliers. Clients involved in this type of project might be the Defence Estates Organisation of the MoD, BAA and London Underground, and typical SCM tools and techniques suggested by Croner (1999) include:
  ■ clustering of suppliers;
  ■ the use of forward planning and demand techniques;
  ■ agile and flexible supply agreements normally using some type of 'framework agreement' or 'call-off' contract arrangements using schedules of rates and partnering philosophies;
  ■ the use of the learning organisation philosophy, and supplier innovation, benchmarking and continuous improvement.

The impact of promoter and demand heterogeneity is consolidated in Table 15.1. The next section explores the concept of the project value chain.

## 15.6 The project value chain

Value chain activities are the basic building blocks from which an organisation creates value for the customers of products or services. The project value chain forms part of an organisation's value chain, since project activities are superimposed on the organisation's normal operating activities. This leads to the concept that a project adds value to the organisation through its own processes. The project value chain concept, which was developed by Standing (1999) is set out in Figure 15.3. There are two primary transition points in the project value chain. The first is the decision to sanction the engineering, and the second is the handover of the completed facility into the operational domain. There are also other transitional points as different organisations become involved. Discontinuities can occur, resulting from changes in values due to the influence of the organisations involved and a different focus is being applied to the project.

Standing's (1999) project value chain framework (Figure 15.4) is sub-divided into three distinct value systems:

- the promoter value system that creates the demand chain;
- the multi-value system, involving parts of the demand chain and the main contractor's supply chain;
- the user value system.

The promoter value chain is concerned with a project to be constructed to meet a business objective, or perhaps a social objective, or a combination of both, depending on the type of promoter. The decision-to-build stage is the point at which the promoter effectively out-sources the 'business project' to the construction industry in the form of a 'technical project' to meet that need. The problem then becomes one of ensuring the alignment of the different organisational value chains involved in the project process to form a holistic value-driven project-focused demand chain working for the benefit of the promoter. For example, a more generic arrangement is set out in Figure 15.5, indicating the levels of complexity that can creep into the project value chain as a supply network. The next section looks at the impact of the procurement system on the project value chain.

**Table 15.1** Promoter and demand impacts on the supply chain.

| | Promoter type | | | | | | | | | |
| --- | --- | --- | --- | --- | --- | --- | --- | --- | --- | --- |
| | Private sector | | | | | | Public sector | | | |
| | Knowledgeable | | | Less knowledgeable | | | Knowledgeable | | Less knowledgeable | |
| SCM demand response | Consumer promoters: large owner occupiers | Consumer promoters: small owner occupier | Speculative developers | Consumer promoters: large owner occupier | Consumer promoters: small owner occupier | Speculative developer | Consumer promoters: large owner occupier | Consumer promoters: small owner occupier | Consumer promoters: large owner occupier | Consumer promoters: small owner occupier |
| Unique | | | ✓ | | | NA | | | NA | NA |
| Customised | | ✓ | ✓ | ✓ | ✓ | NA | | | NA | NA |
| Process | ✓ | ✓ | | | | NA | ✓ | ✓ | NA | NA |
| Portfolio | ✓ | | ✓ | | | NA | ✓ | ✓ | NA | NA |

*Note:* a ✓ denotes that this is the probable occurrence; NA indicates no occurrence; a blank indicates a possible but unlikely occurence.

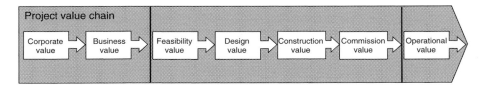

**Figure 15.3** The project value chain. Source: Standing, 1999, Figure 8.10.

## 15.7 Procurement and the project value chain

Provided that the project value chain remains in alignment, it is a series of inputs and outputs that create value for the promoter. Each value transition should be adding value until the complete project forms an asset for the promoter's organisation to meet a corporate need. Complexity is added to the project value chain when other value systems impart skills and knowledge to it, or create barriers to its effective operation as an integrated system on behalf of the promoter. One of the most important strategic decisions that has an impact on the project value chain is the choice of procurement route, which can act as either an enhancer or a barrier to value creation and improvement. Single-point responsibility for the whole delivery from concept to operation for the promoter comes to the fore and should, in theory, create the capability to maintain the integrity and alignment of the project value chain.

Figure 15.6 is a schematic diagram comparing some of the major procurement systems with the project value chain concept. Schematically, the procurement systems at the top of the diagram provide more opportunity to maintain the integrity of the project value chain, since an increased number of discrete activities come under one umbrella organisation for single-point delivery. Whilst the project remains within the promoter value system, value should be maintained internally, although once transfered into the multi-value system through procurement, there is a potential for loss in promoter value.

The contractor-led procurement systems, where the contractor offers a one-stop-shop service, to a greater or lesser extent have the potential for greater integration within the project value chain, depending on the method of tender. Profession-led design procurement systems involve additional interfaces and provide more opportunities for disruption to the project value chain. Under profession-led systems, any change by the contractor must have the promoter's approval, since they are the only party that can sanction change. No system has a mechanism which permits the contractor to change the design.

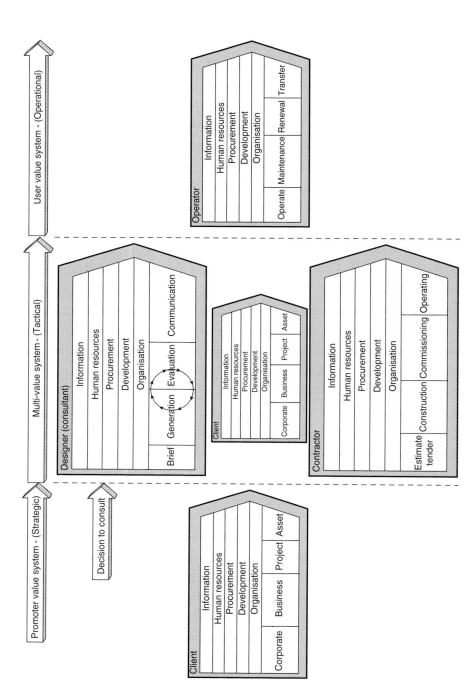

**Figure 15.4** Alignments in the project value chain (adapted from Porter). Source: Standing, 1999, Figure 8.1.

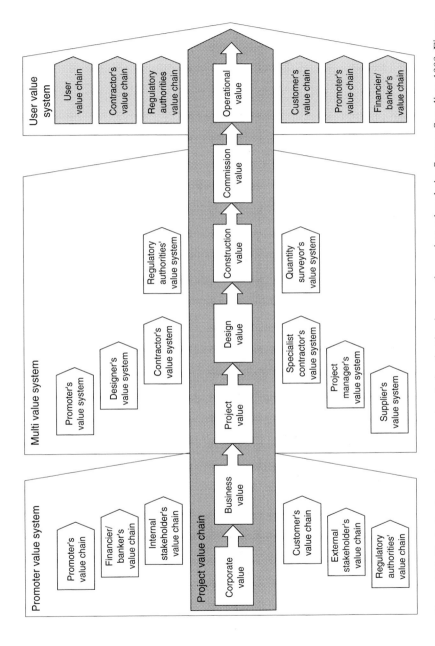

**Figure 15.5** Typical value systems and value chains that impinge on the project value chain. Source: Standing, 1999, Figure 8.11.

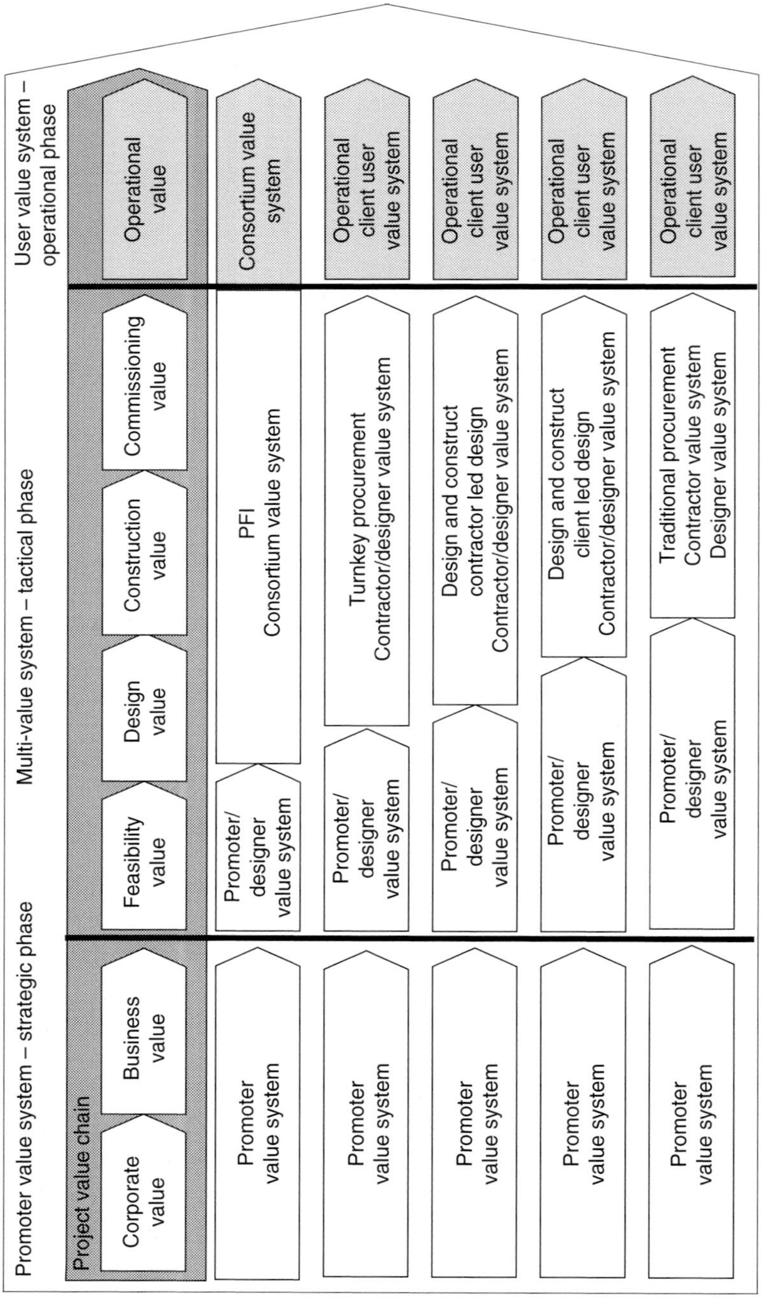

**Figure 15.6** Schematic diagram of procurement systems superimposed over the project value chain. Source: Standing, 1999, Figure 4.7.

The management forms of procurement lie somewhere in the middle of the diagram. They permit increased involvement of engineering knowledge early in the process, but essentially they are profession-led routes with a consequent increase in the number of interfaces. Turnkey procurement has similarities to PFI, but unlike the latter, turnkey procurement does not have the additional requirement and liability for operating the facility. The contractual positioning and role of the designers will alter the impact on the project value chain. If the contractor employs designers in-house, then there should be increased alignment in the project value chain. However, turnkey, where the designer is independent of the contractor, imposes another value system.

The traditional construction procurement route, at the bottom of the diagram, is probably the most disruptive to the project value chain, since single-stage competitive tendering occurs at the transition point between design value and construction value. Two-stage tendering, overlaid onto the traditional procurement route, does, however, have the capability of bringing construction expertise into the project much earlier. Therefore, there is also an interaction between procurement method and choice of tendering strategy in terms of impact on the project value chain.

To summarise, the choice of procurement route is a strategic decision made by the promoter and/or its advisors that has a fundamental impact on the demand chain. It has the capacity to assist or hinder the transfer of value through the project process. The use of value management and value engineering are methodologies for aligning or re-aligning the project value chain.

## 15.8 Prime contracting

Two major factors have caused the MoD to change its approach to procurement. The first is the abandonment of compulsory competitive tendering in the public sector in favour of the best value regime, that came into effect on 1 April 2000. This requires government clients to think in terms of the life-cycle cost of designing, building, operating and maintaining a facility. The second is rethinking construction, or the Egan Report, which builds on the Latham Report and has had a major impact on government thinking at all levels. The government is committed to implementing the principles identified in the Egan report, including partnering, supply chain management and continuous improvement.

The Defence Estates Organisation (DEO) of the Ministry of Defence has taken the initiative with its *building down barriers* project to adopt a radical approach to procurement, and has embraced the principles of supply chain management and strategic alliancing within its new procurement process, which is called *prime contracting*. The prime contractor has single-point responsibility for designing and building, or designing, building, operating and maintaining a facility. The distinguishing characteristic of prime contracting from other procurement routes is seen as *the* requirement for single-point responsibility for the total process, from concept to operation. There are, however, some similarities with projects delivered under PFI, which also provides single-point responsibility. The primary differences are that no financing is required by the prime contractor, and there is a limit to the maintenance and operating period.

The prime contracting initiative was established in January 1997 and ran until late 2000. It was jointly funded by the then Department of the Environment, Transport and the Regions (DETR), the DEO and the MoD. The initiative had three primary objectives:

(1)   to develop a new approach to construction procurement, i.e. prime contracting, based on supply chain integration;
(2)   to demonstrate the benefits of prime contracting in terms of improved value for the client and profitability for the supply chain through two pilot projects;
(3)   to assess the relevance of the new approach to the wider UK construction industry.

There are two main themes to the building down barriers project. One is undertaking pilot projects using the new procurement route, and the other is a research and development theme involving the development of the new procurement process and a 'tool-kit' to support it, and then evaluating the pilots and the tool-kit.

The DEO, in conjunction with the Tavistock Institute, the Warwick Manufacturing Group (WMG), Amec Construction, John Laing Construction, British Aerospace Construction Consultancy Services and the Building Performance Group, set up two pilot projects where prime contractors were given full responsibility for the design, construction and maintenance for a 'proving period', including an additional focus on operational and through-life costs. Consultant and trade suppliers became part of the prime contractor's supply chain. The intention is that the MoD's £2bn per annum procurement regime will be let under the prime contracting route, whereby the number of supplier firms will contract by approximately 90%, and the value of its main construction contracts will increase to £100–200 m, with benefits accruing in the longer term across multiple projects.

Prime contracting has already come under close scrutiny by architects, who have expressed concerns that the prime contracting route will further undermine their professional standing, since the drive will be towards repeat solutions. It has also been viewed as institutionalised design and build.

There are five phases in the whole life of a prime contracting project. These are set out in Figure 15.7. The core conditions envisage prime contracts being divided into two main categories: first, the 'one-stop-shop', where the contract period can range between 5 and 10 years, and second, major stand-alone capital works, where the prime contractor will design and construct an asset with a through-life compliance period of approximately 3 years.

Both prime contracts are based on the core principles listed below.

- Using supply chain management as a fundamental underpinning to prime contracting.
- A commitment to collaborative working.
- Use of open-book accounting.
- Fraud prevention and detection policies adopted by the prime contractor.
- The use of an integrated project team (IPT) as a critical success factor. The IPT initially starts with the promoter, and then incorporates the prime contractor and the supply chain when they are appointed.
- Seeking innovative solutions to improve value-for-money and continuous improvement.
- Use of output specifications. For one-stop-shops, the range of services will include:
  - capital works, including refurbishment;
  - planned maintenance;
  - reactive maintenance;
  - soft facilities management services, i.e. cleaning, catering, space planning and others, depending on the type of prime contract.
- The use of best practice in value engineering and through-life costing as mechanisms to provide better value for money as well as a more transparent view of the cost of ownership.
- Warranting fitness-for-purpose for all design and construction work for the intended purpose, as set out in the strategic brief.
- The use of a maximum price target cost arrangement, including procedures for sharing 'pain and gain'.
- Transfer of undertakings from previous employer (TUPE) arrangements using the MoD's Code of Practice.

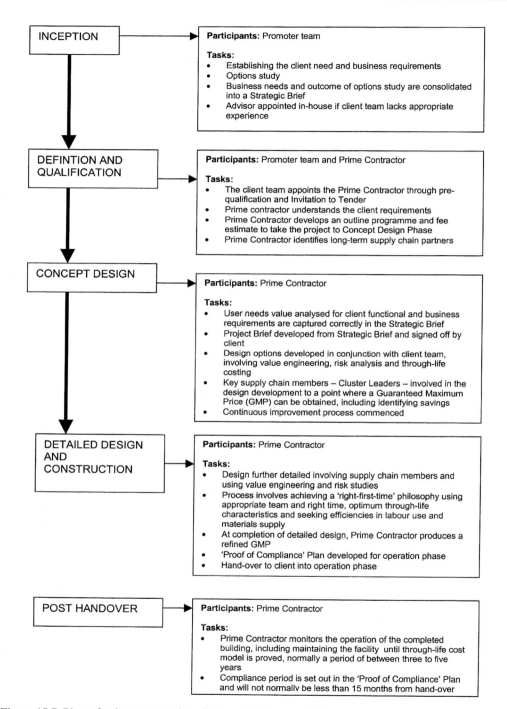

**Figure 15.7** Plan of prime contracting. Source: Holti *et al.*, 1999.

❑  The use of a dispute resolutions board.

## 15.9  The operation of future construction supply chains

In future, it is likely that the construction supply chain will be much wider than is currently envisaged, and will include, on a regular basis, funders, designers and other members of the design team, a 'general' contractor, specialist contractors and suppliers of components, providers of services and facilities managers. They will offer and deliver a total service package to the promoter, from concept through to use. Extrapolating from current trends in the industry associated with design and build and more importantly, PFI and prime contracting, the optimum solutions project by Brown *et al.* (2000) for improving the delivery of multi-storey steel-framed buildings envisages that *strategic supply chain brokers* and *phased, seamless teams* will emerge in the industry and become the norm. The emergence of, and pressures for, the strategic supply chain broker and phased, seamless teams are outlined below.

### *The emergence of the strategic supply chain broker role*

The emergence of the role of strategic supply chain broker is seen as a natural, direct consequence of the UK construction industry having a long and engrained record and culture of adversarial contracting. Through initiatives already mentioned, such as the Latham report, the Egan report, M4I and the Construction Best Practice Programme, there is a clear determination by policy makers that things have to change. The globalisation of markets, especially in the service sector, coupled with governments' strategies of public sector privatisation in many countries, will also increase the pressure for change and attract investment and foreign firms to deliver new types of combined public/private sector projects. The concept of the strategic supply chain broker directly assails the adversarial culture of the UK construction industry, and the need for integrated supply chain management argued for in numerous government initiatives. Regular procuring clients are already pressurising the industry for a 'one-stop shop' service, coupled with cost and time certainty in delivery, and they are very vociferous in their demands for the industry to meet their expectations as customers.

Newer procurement routes, such as PFI and Prime contracting, as well as more integrated, team-based routes, such as management contracting and construction management, are attempts to draw together design and construction interfaces and responsibilities. PFI and prime contracting also incorporate the operational phase. The broker role is therefore an organic extension of these procurement options and industry-led initiatives such as that pioneered by MACE with the launch of their 'branded product'. The branded product is a response to rethinking construction using the core philosophy of a fully integrated design and delivery process that ensures that value, quality and speed of delivery are 'designed in' from the outset. Use is made of standardised pre-designed buildings and components, allowing engineering solutions, systems and architectural component interfaces to be identified and resolved early. In addition, at the heart of the branded product are IT/CAD systems that allow work to proceed concurrently. PFI, prime contracting and developments such as MACE's branded product are precursors to the emergence of the broker role, i.e. one that will take responsibility for the contracted-out strategic, tactical and potentially the operational delivery of an asset for a promoter. The key is that the broker is able and willing to take the risk away from the promoter as a one-stop-shop supplier of skills, expertise, product components and knowledge, for a price. Optimum solutions research suggests that brokers will compete with their expertise using their supply chain base.

In the industry of the future, it is predicted that two main forms of broker would emerge, the *volume broker* and the *innovative broker*, based on the type of demand that will exist in the industry, and the requirements of knowledgeable and less knowledgeable promoter. Using the Pareto principle, approximately 80% of all projects can be categorised as 'routine', and would act as one of the drivers for working with the same project teams. This would form the basis of volume delivery within the industry, and volume brokers would emerge over time who would develop a brand reputation for timely, regular delivery to cost at an appropriate quality and functionality. Approximately 20% of projects can be categorised as technically demanding, innovative or leading-edge. The current management forms of procurement would be the drivers behind team delivery for this type of project. Brokers emerging from this domain would develop brand reputations as 'innovative brokers'. The requirement here would be for the phased seamless team to be put together to deliver innovative solutions. Brokers operating under these conditions would have a pool of leading-edge supply chain members who would be niche providers to the industry.

### The phased, seamless team

Contractually based relationships highlight the dangers of adversarial interactions developing when things go wrong. They also tend to be much more linear, with one task or sequence of events following another, with different organisational relationships procured at a particular time to suit a particular project programme. This may militate against securing the right skills at the right time, regardless of the position within the supply chain. Underpinning the operation of the supply chain broker and the construction supply chain is the *phased, seamless team*. Seamless teams would pre-exist around a particular supply chain broker. Members of these teams, through long-term working relationships, would be familiar with each other's working practices, including a full appreciation of the requirements for appropriate and timely information exchange to allow key project interfaces to operate effectively throughout the supply chain. The seamless team would operate on the basis of a moral and psychological contract founded on trust. Its operations will not be contractually based, and rewards and incentives would be based on performance indicators tied into promoter-focused value and not cost-driven service delivery. Such teams would, through common goals, operate in a cooperative, solutions-oriented culture where pooling and sharing of information would occur through a combination of face-to-face problem-solving meetings supported by information technology. As the project progresses, the team would operate with seamless handover processes and procedures, as some new members enter and others leave the team owing to changing task requirements. To support this, expectations, roles, responsibilities, skills and competencies would be known and fully understood among team members. Under a system where a psychological and moral 'contract' based on trust is the founding principle of a relationship, deeper levels of commitment can occur, and the principles behind the seamless team would permit concurrency of inputs and skills for the benefit of the project and product, and buildability and decision-making to take account of all aspects of the total process. Interfaces would be owned, and information flows improved, since they are internal to the team. This would lead to savings in both time and money. The seamless team would also provide greater opportunities for more continuity of work and learning.

## 15.10 Summary

This chapter has suggested that the supply chain system in engineering comprises two components, the demand chains of numerous promoters, generated through the procurement route adopted, and the supply chain of the main contractor that has to respond to a diverse range of promoter types, and hence demand chains. Supply chain management is now on the strategic agenda of firms, and is no longer seen as an operational issue. It involves thinking about activities within the supply and purchasing function that are internal and external to the firm. Supply chains in engineering can be classified as under three main headings.

❏ *Professional services.* These suppliers provide a combination of skills and intellectual property, typically comprising the designers and other professional consultants. Those supplying professional services operate under numerous professional institutional umbrellas. In this chapter, the delivery of professional services, a tertiary-level provision, falls within the domain of either the promoter's project-focused demand chain or the main contractor's supply chain depending on the procurement route adopted.
❏ *Implementation and assembly.* This on-site construction process is highly skills-focused and also forms part of the service sector of the

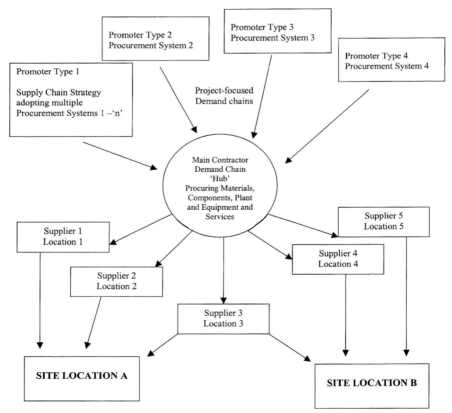

**Figure 15.8** Engineering supply chain system.

industry. It will comprise a range of different types of contractors either with overall responsibility for the management of the process, or for supplying inputs to the process. It involves fitting, installation, assembly, repair, and on-site and off-site labouring. In this chapter it is also viewed as a manufacturing process.

❑ *Materials and products.* This comprises the materials, products and hired plant involved in the on-site process. These form part of the main contractor's supply chain, and will involve the make, move and store activities to and at a particular site.

Figure 15.8 is a schematic diagram of the supply chain system in engineering, from which supply chain strategy is developed by the main contractor or, as research described in this chapter has indicated, also by clients to the industry.

## References

Brown, D., Williams, P., Gordon, R. and Male, S.P. (2000) *Optimum Solutions for Multi-Storey Buildings*, Final Report to the DETR.

*Croner's Management of Construction Projects*, July 1999, pp. 2–541.

Holti, R., Nicolini, D. and Smalley, M. (1999) *Building Down Barriers – The Concept Phase*, Interim Evaluation Report, March, p. 1. http://www.coconet/supplychain/prime-principles/prime-principles.html

Saunders, M. (1997) *Stategic Purchasing and Supply Chain Management*, 2nd edn, Financial Times and Pitman Publishing, London.

Standing, N. (1999) *Value Engineering and the Contractor*, unpublished PhD thesis, University of Leeds.

## Further reading

Aitken, J. (1998) *Supply Chain Integration within the Context of a Supplier Association*, unpublished PhD thesis, Cranfield University.

Christopher, M. (1998) *Logistics and Supply Chain Management: Strategies for Reducing Cost and Improving Service*, 2nd edn, Financial Times and Pitman Publishing, London.

*Croner's Management of Construction Projects*, Special Report No. 11, 17 May, p. 3.

Egan, Sir J. (1998) *The Egan Report. Rethinking Construction*, DETR Publications, Rotherham.

Gatorna, J.L. and Walters, D.W. (1996) *Managing the Supply Chain: A Strategic Perspective*, Macmillan Business, Basingstoke.

# Chapter 16

# Team-Based Supply Chains and Partnering

This chapter examines the impact of team-based supply chains and partnering on project management performance. It will identify which internal processes need attention to make real performance improvements. The key areas of alignment, continuous improvement and issue resolution will be addressed, so that readers can identify actions within their own project management environments that would lead to partnering.

## 16.1 Background

Partnering has been shown to yield significant savings in time and cost. Longer-term strategic partnering arrangements have been shown to yield the greatest benefits, with cost savings of up to 40%. Sir Michael Latham, in his report 'Constructing the Team' (1994), recognised the role that partnering could play in reducing conflict and improving efficiency in the construction industry within the UK, and recommended its use. Since then, a number of partnering arrangements have been entered into in the UK, and their performance has been analysed and compared with alternative contractual relationships. It is clear that there are a number of advantages to be gained from partnering, but there are also problems that need to be overcome.

Commitment and clear leadership by example and involvement from senior management are required for team working and partnering to succeed. It is the enthusiasm, commitment and desire for a successful outcome by every individual involved in the project that contributes to its success. Change and behavioural management specialists can help to accelerate the cultural change programme.

Partnering arrangements, whether project-specific or longer term, do not replace the need for competition at the outset of the project. This competition should be on the basis of value for money over the life of the facility, and not on cost alone. It will need to address each organisation's culture and commitment to work in a collaborative manner to drive down unnecessary costs whilst delivering the required quality.

Incentives matched to shared objectives should be included in partnering arrangements to encourage the whole team to provide benefits for the promoter significantly beyond those contracted for (e.g. by using innovation or different working practices to deliver the same or better services while yielding cost savings). Incentives should not just be awarded for doing a good job.

## 16.2  Team working

Team working, partnering and incentives are not soft options that create a cosy environment in which to work. They are management techniques that deliver the aspects that are of value to the promoter by eliminating or minimising wasteful activities not directly contributing towards those aspects of value. They require leadership, commitment and continued involvement from the top down.

Ultimately, the success of a project comes down to people. It is the enthusiasm, commitment and desire to succeed that will determine the outcome of the project. Particular attributes required of individuals include:

❏  continuously searching for improvement;
❏  actively listening to, and learning from, others;
❏  identifying which personal attributes we need to improve, to allow us to work better with others.

When selecting a project team, the personal attributes of the individuals should be taken into account. Where it appears that, even with training, an individual is unlikely to be able to work constructively as part of the team, that person should not be selected. To prevent confusion and misunderstanding, the roles and responsibilities of each individual should be clearly set down at the outset.

Team working is defined as working together as a team for the mutual benefit of all by achieving the common goals whilst minimising wasteful activities and duplication of effort. When problems or difficulties are encountered, all parties should work together as a team to overcome those problems rather than blame each other. Team working is the starting point on which relationships with others should be built, and is equally important internally as externally. It does not replace proper and appropriate management structures, but is a pragmatic way of working together to meet the project objectives. It should encourage greater openness, and the earlier involvement of contractors and suppliers.

The benefits of team working include:

❑   understanding each other's objectives;
❑   use of collective brain power to find solutions;
❑   reduced numbers of personnel to monitor progress and prepare claims;
❑   reduction in correspondence;
❑   reduction in the duplication of effort;
❑   elimination of 'man to man' marking;
❑   improved working environment, i.e. cooperation not conflict;
❑   enhanced reputation of the individual when associated with successful projects;
❑   reduction in issue escalation.

A team-working culture can be encouraged by holding team-working workshops and by arranging for team members to undergo team-building training. Co-locating is also believed to have a positive impact on team working. Where that is not practicable, attention needs to be given to establishing a common virtual environment through intranets and other such mechanisms.

## 16.3  Partnering

Partnering extends the definition of team working by adding the need for a more formal structure to be agreed by the parties which:

❑   identifies common goals for success;
❑   sets out an issue-resolution process;
❑   identifies the targets that provide continuous measurable improvements in performance;
❑   sets out gain and pain share arrangements (incentives).

These elements are normally set down in a partnering charter. However, the partnering arrangement and charter do not replace the need for a formal contract. Although there may be a number of standard contracts between promoter and suppliers/contractors there should be a single partnering charter.

There are many definitions of partnering, some of which are given below. Partnering was defined by the Reading Construction Forum as:

> ... a management approach used by two or more organisations to achieve specific business objectives by maximising the effectiveness of each participant's resources. The approach is based on mutual objectives, an agreed method of problem resolution and an active search for continuous measurable improvements. (Bennet and Jayes, 1995)

The US Army Corps of Engineers (1991) defined partnering as:

> ... the creation of an owner–contractor relationship that promotes the achievement of mutually beneficial goals. It involves an agreement in principle to share the risks involved in completing the project and to establish and promote a nurturing partnership environment.

The Construction Industry Institute Partnering Task Force (1996) defined partnering as:

> A long-term commitment between two or more organisations for the purpose of achieving specific business objectives by maximising the effectiveness of each participant's resources. The relationship is based on trust, dedication to common goals and an understanding of each other's individual expectations and values. Expected benefits include improved efficiency and cost-effectiveness, increased opportunity for innovation and the continuous improvement of quality products and services.

According to the Associated General Contractors of America (1991) partnering is:

> ... a way of achieving an optimum relationship between a customer and a supplier. It is a method of doing business in which a person's word is his or her bond and where people accept responsibility for their actions. Partnering is not a business contract, but recognition that every business contract includes an implied covenant of good faith.

The Egan Report (Egan, 1998) states that:

> Partnering involves two or more organisations working together to improve performance through agreeing mutual objectives, devising a

way for resolving disputes and committing themselves to continuous improvement, measuring progress and sharing the gains.

As a concept, partnering is very difficult to capture in a short definition and that is why a number are presented. The Egan Report definition does identify the key elements that must be present for partnering to exist, and is the one the author usually recommends.

From the first handshake to the finished product, partnering provides the construction industry with a fundamentally different approach to teamwork. This revolutionary management process emphasises cooperation rather than confrontation, and it calls on the simple philosophy of trust, respect and long-term relationships.

Partnering should not be confused with other good project management practices, or long-standing relationships, negotiated contracts or preferred supplier arrangements, all of which lack the structure and objective measures that must support a partnering relationship. These other forms of team-based supply chain arrangements are not being ignored, but rather, it is suggested that the adoption of partnering proper will lead to greater benefits.

The main difference between partnering and other forms of procurement is the shift in the commercial route by which projects are procured. By changing the procurement and completion process, the parties can reap the benefits of cost saving, profit sharing, quality enhancement and time management. In the search for partnering relationships, promoters are gradually altering their procurement criteria. There is a shift from 'hard issues', such as price and the scope of work, towards 'softer issues' that revolve around attitude, culture, commitment and capability.

The key to partnering is that it starts at the outset of a project, and is applied to the selection of the design professionals, consultants and contractors/suppliers. Partnering arrangements are flexible, and may be formal or informal depending on the size, complexity and scope of the project. What is essential, however, is an understanding among all parties involved that they are committed to a plan to anticipate problems and deal with any as they arise. Figure 16.1 illustrates this process.

Partnering can be either 'project-specific' or 'strategic'.

❑ *Project specific partnering:* A method of applying project-specific management in the planning, design and construction profession without the need for unnecessary, excessive or debilitating external party involvement.

❑ *Strategic partnering:* A formal partnering relationship that is designed to enhance the success of multi-project experiences on a

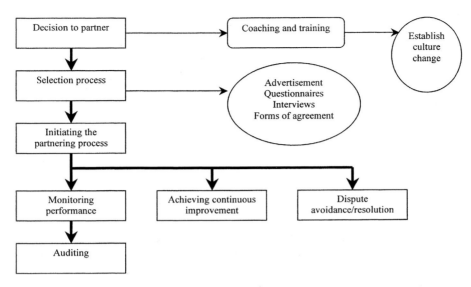

**Figure 16.1** The partnering process. Source: ECI, 1997.

long-term basis. Just as each individual project must be maintained, a strategic partnership must also be maintained by a periodic review of all projects currently being performed.

There are three essential features that characterise a partnering relationship, and without these the relationship is unlikely to be successful. These features can also be used by organisations that are not partnering, but the full benefits will not be realised if the practices are used in isolation. Incentives are an important tool in the achievement of the mutual objectives.

*Mutual objectives*

Despite a widespread belief that these are obvious to all involved, team members typically do not understand other members' objectives very well at the beginning of the project. Developing a shared understanding of the mission and clear, detailed objectives is essential to a successful project. By getting people together to clarify what is important to each of them, and what their expectations for themselves and each other are, many suspicions, false expectations and lurking issues are dissipated. The agreed mutual objectives must constantly be kept under review through meetings and effective communication. The principle that adversarial attitudes waste time and money must be mutually accepted.

Partnering is a tried and effective technique for placing the responsibility for anticipating problems early on those best equipped to effect their resolution. The partnering concept has been proved to be a powerful participant-driven method of anticipating problems that might be encountered on a project.

### Issue avoidance

The primary focus is on the prevention or early resolution of issues. Partnering is seen as a method of settling conflict and disputes before they become destructive and costly. Early intervention is critical, because as issues develop into disputes, positions become hardened, formal claims are made and costs go up. In partnering, the issue resolution process is formally designed in 'kick-off' sessions between the partnering members so that everyone has a clear understanding of what the process is and agrees to use it. It is based on seeking 'win–win' solutions and not parties to blame. The issue resolution process illustrated in Figure 16.2 should be followed for a speedy settlement.

### Continuous and measured improvement

This on-going process involves a fundamental understanding of customers' requirements, identification and aim for best practices, development of new outputs and processes, implementation of new changes, and the evaluation of the effectiveness of the improved process. The evaluation and feedback measures not only ensure that the objectives of the project are met, but that they also aid in planning future projects. They identify and quantify what worked, what did not, and how things could be improved next time. Measurement also keeps the partnering team focused on the terms of the partnering agreement.

### Incentives

Incentives should encourage the parties to work together to eliminate wasteful activities that do not add value to the promoter, and to identify and implement process improvements, alternative designs, working methods and other activities that result in added value. Incentives should not be given for merely doing a good job; nor should they be given for improvements in performance that are of no value to the promoter. Incentives can be built into the bidding contract or into the non-binding partnering charter. Performance targets on which incentives are based must be measurable, and they should cover not only cost and time, but also safety and quality.

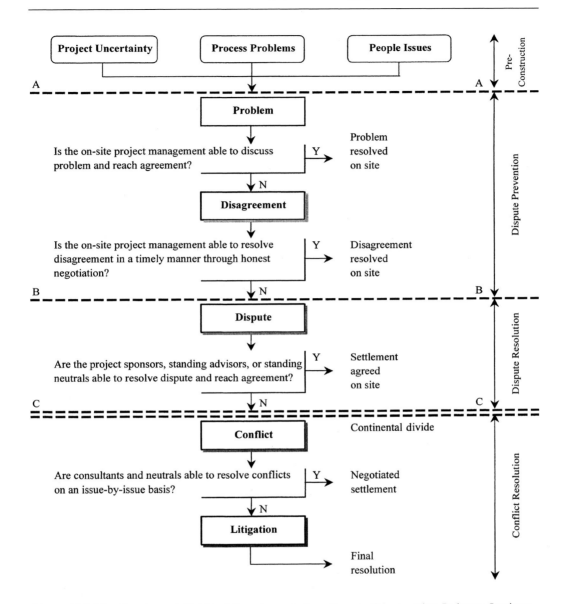

**Figure 16.2** The Continental divide of dispute resolution. Source: Construction Industry Institute (CII), 1995.

## 16.4  Establishing the relationship

Partnering is generally established through a structured, facilitated process, normally consisting of organising workshops to bring the participants together. The workshop is designed to provide a positive partnering atmosphere. Its purpose is to establish commitment, trust, and mutual goals and objectives among all the members of the partnership. The length of the workshop will depend on such variables as the complexity of the contract, the experience of the participants in partnering, the number of partners, and the time needed for team-building.

In most cases, a facilitator-directed partnering workshop will accelerate the successful implementation of the partnering effort. The facilitator is a neutral person who helps the partners to get organised from the outset of the process. The facilitator helps develop the partnering workshop and also leads it, and is instrumental in encouraging the parties to design their charter, identify potential problems and develop a conflict escalation procedure. The facilitator may also deal with any scepticism or bias brought to the workshop, and keep the team focused on the partnering process. Keeping the facilitator involved maximises the benefits to the partners by keeping them on the partnering path, and by increasing the facilitator's knowledge of the programme, contract requirements and unique contract administration issues. The workshop may entail both individual and joint sessions with the facilitator, and will generally last at least 2 days. During the workshop, the essential ingredients of the partnering arrangement are drafted.

❏  The roles and responsibilities for each partnering participant.
❏  Specific programme issues and concerns, with an action plan being developed for each.
❏  Conflict management techniques and specific procedures to resolve conflicts among stakeholders.
❏  An alternative dispute resolution (ADR) approach.
❏  Metrics for the assessment of accomplishments.
❏  Techniques designed to improve team communications.
❏  The partnering charter.

At the end of the initial partnering workshop, the participants must identify the goals, develop a plan to achieve them, and draw up a partnered agreement, or charter, to make a commitment to them. The partnering workshop is just a starting point. Periodic reviews at regular intervals are critical to success. These reviews may involve an assessment of the partnering relationship and follow-up workshops. One reason why it is beneficial to keep the facilitator informed during contract performance is to enhance their involvement in follow-up workshops if they are required. Follow-up workshops should also be considered when major players in the partnering process are replaced, in order to ensure that the new participants are knowledgeable about, and committed to, the process, and if there is a breach of the charter or conflict escalation procedure, or some other indication that it is necessary to reaffirm the process and remind participants of the need for their consistent commitment.

The partnering charter is the focal point of the relationship and the blueprint for success. There is no single approach to drafting a partnering charter. Some of the general matters that should be addressed in the charter, according to the European Construction Institute (ECI), are listed below.

- ❑ A statement committing the parties to abide by the aims of partnering.
- ❑ An expression of intent to communicate freely.
- ❑ Acknowledgement that problems may occur during the currency of the contract.
- ❑ Acknowledgement that relationships may deteriorate due to disputes being left unresolved.
- ❑ Commitment to monitoring the partnering process performance in accordance with the procedures set up at the workshop.
- ❑ Commitment to a safe project with a high-quality outcome.

While not legally binding, the charter is a 'social contract' about how the team members have agreed to work together. The charter constitutes the first visual instrument of shared partnering commitment. It states each goal in specific terms, which are measured periodically. During the course of a project, joint evaluations of the charter keep its original objectives in place and underscore their continuing importance.

The charter is not intended to be a contractual document; nor does it supersede the contract. The contract establishes the legal relationship between the parties, while the charter is concerned with the working relationships. It is, in effect, a statement of how the parties intend to conduct themselves. The charter represents the common commitment of all the participants to the partnering process.

The partnering contract should not be allowed to become a barrier to the participants. From a purist standpoint, the contract is not an important part of the partnering process. Nevertheless, a formal agreement needs to be established between all parties. However, the contract should promote and complement the partnering process, and not contribute to an adversarial relationship. The success of the contract as a working document depends on the commitment to fulfill the mission statement and the charter supporting that statement, as well as a statement of expectations and objectives. Flexibility should also be built into the contract to allow for change that may be a result of the continuous improvement or review process.

## 16.5  Making the relationship work

When entering a partnering agreement, there are a number of contractual issues that need to be considered.

❑ What effect established benchmarking will have upon the improvements striven for by the parties.
❑ What redress an injured party can claim if the agreement is breached.
❑ Whether it provides recompense in the event that either party invokes the termination provisions.
❑ Whether there are any guarantees that the consultants and contractors will be appointed for any or all of the individual projects.

No matter how committed management and team participants are, the partnership will not run itself. To track, care for and feed the process one individual designated by each partner must assume responsibility. This 'champion' will play a powerful and influential role in the process, and will generally be at the project management level. They will oversee the project, reinforce the team approach, overcome resisting forces, participate in the resolution of issues escalated to their level, celebrate successes and maintain a positive image for the project. They will provide the leadership necessary to ensure that the partnering process moves smoothly throughout the performance of the contract. They will also communicate with senior management officials to keep them apprised of partnering efforts and to solicit their continuing commitment. Champions are more than figureheads. They must play a vital role in initiating and energising the partnering process for those on the team, and implementing the tools developed at the partnering workshop.

## 16.6  Benefits of partnering

Supporters of partnering advocate its use by suggesting benefits such as improved efficiency, reduced costs, reliable quality, faster construction, completion on time, continuity of work, sharing of risk, reliable flow of design information and lower legal costs. The benefits could be summarised as:

- improving cooperation between design and implementation teams, giving rise to a product more suited to promoter needs;
- motivating innovation;
- increasing the willingness to solve design and site problems, which will, in turn, reduce delays and inefficiencies;
- encouraging the sharing out of identified savings in time and cost;
- reducing potential claims;
- encouraging good service and improving subcontract quality and timeliness;
- speeding up of decision making;
- establishing a relationship between parties that may lead to future work.

Much of the interest surrounding partnering is centred on improvements in efficiency and profitability and reductions in costs. Various studies have been undertaken on partnering arrangements. Major benefits reported by the US Army Corps of Engineers (1991) include:

- a reduction in overall project costs;
- a reduction in promoter's costs;
- a reduction in time taken to reach completion;
- lower risk of cost overruns and delays because of better time and cost control over the project;
- a virtual elimination of time overruns, and an increased potential to expedite the project through efficient implementation of the contract;
- a reduction in paperwork and administrative costs;
- an increase in contractor's profits;
- significant improvements in site safety;
- improvements in morale amongst the project team;
- reduction of disputes and formal claims, and projects are completed with no outstanding claims or litigation;
- improvements in communication;
- higher trust levels.

The Reading Construction Forum (Bennet and Jayes, 1995) suggested improvements in cost savings of 2–10% for project-specific partnering, and of 30%, over time, with strategic partnering, in their report on partnering arrangements both in the UK and abroad. Such benefits appear to stem from the atmosphere of mutual trust and collaboration which exists within a partnering environment. The contractor is able to develop a special relationship with the promoter which is similar to that of a design and build contract.

Partnering provides the appropriate environment to discuss the client's requirements and to develop strategies for achieving them. The contractor, meanwhile, can offer savings in costs due to reductions in abortive tendering made possible by the price negotiation arrangement. This can be extended to subcontractors as well. Subcontractors may be a part of the partnering arrangement or, alternatively, can provide the contractor with a reduced price for 'guaranteed' work, in a similar way to that utilised in serial contracting. Equity involvement improves the subcontractors' opportunity for innovation and value engineering. Involvement in decision making can avoid costly claims. Additional benefits to the subcontractors include an improved cash flow, reduced overhead costs and greater profit potential, better and more reliable programming, and enhanced opportunities for repeat business.

Further savings may accrue from a reduction in variations. This is a direct result of a better understanding, by all concerned in the partnering arrangement, of the promoter's needs from the early stages of the design process. The design process itself can also benefit from the contractor's contribution. Efficiency can be considerably improved during the construction phase by focusing on buildability issues. By replacing the potentially adversarial atmosphere of a traditional owner–constructor–engineer relationship with an environment of trust and enhanced communication to achieve a set of common goals, project quality, safety and productivity can be improved. An effective partnering agreement can show considerable reductions in the amount of abortive work undertaken. It is also clear that greater savings are likely to spring from a long-term alliance than might be possible where the partnering arrangement only lasts for one project. There are bound to be improvements in working methods and communication systems resulting from the evolutionary development of the partnering relationship.

Bringing the team on board early makes it possible to concentrate effort at the beginning of the project. There are also benefits to be gained from the use of participative management techniques in partnering arrangements. When a team gets together to set common goals, it establishes mutual trust and a commitment to achieve those goals. It is not surprising, then, that once the project is underway, pressures will be self-imposed by each member of the team to achieve the targets that they have helped to set.

Through a partnering arrangement, the promoter and contractor, normally adversaries in the construction process, are prepared to deal with situations together, both promptly and effectively. While the partnering concept was developed by promoters, even contractors are beginning to appreciate the reduced exposure to litigation through open communication and issue resolution strategies. The commitment to resolve disputes informally at the earliest opportunity minimises the necessity for litigation in administrative and judicial forums. Avoiding the considerable expense and delay attributable to litigation frees the partnering participants to concentrate their efforts on successful and timely contract performance. At the outset of the relationship, the parties determine how they will manage any conflicts that might arise. This is often accomplished through a conflict escalation procedure. This procedure identifies the roles and responsibilities of individuals from both client and contractor, and provides for the automatic elevation of issues through several organisational levels to avoid inaction and personality conflicts.

Partnering focuses on the mutual interests of the parties. Rather than the parties developing individual positions on issues, partnering engenders a team-based approach to issue identification and problem resolution, which is focused upon the accomplishment of the parties' mutual objectives. Despite the potential advantages and benefits that can be derived from a partnering relationship, partnering also entails risks that should be evaluated.

## 16.7  Constraints to partnering

Perhaps the greatest concern regarding this method of working relates to the need for openness, trust and commitment by all the parties concerned. Contractual relationships within the construction industry have been founded on mistrust and adversarial practices, which are not easy to discard in favour of a more open-minded approach. 'Win–win' thinking is an essential element for success in this process. It has been estimated that in a traditional contract arrangement, approximately 30% of the total project cost can be attributed to the employment of layers of additional, often unnecessary, staff. Therefore, a new partnering culture that will replace the old adversarial practices has to be developed.

In her studies of partnerships in many types of organisations, Rosabeth Moss Kanter (quoted in Moore *et al.*, 1992) observed the difficulties inherent in achieving successful partnering. In her view 'The fragility of interorganisational alliances stems from a set of common "dealbusters" – vulnerabilities that threaten the relationship. Partnerships are dynamic entities, even more so than single corporations, because of the complexity of the interests in forming them. A partnership evolves; its parameters are never completely clear at first, nor do partners want to commit fully until trust has been established. And trust takes time to develop. It is only as events unfold that partners become aware of all the ramifications and implications of their involvement.'

Moore *et al.* (1992) reported four major 'roadblocks' to success.

❏ *A shift in business conditions.* If conditions change and the project is behind schedule, with unanticipated technical problems and cost overruns, the strategy within each organisation may change and even revert to an 'us versus them' attitude.

❏ *Uneven levels of commitment.* Unevenness of commitment often develops from the basic differences between organisations.

❏ *Lack of momentum.* A partnership requires nurturing and development throughout the life of the project. The representatives from each side must constantly work to maintain the health of the partnership.

❏ *Failure to share information.* Partnering requires timely communication of information and the maintenance of open and direct lines of communication among all members of the partnering team. The failure to share information is most likely to arise when team members revert to past practices.

People often prefer to work as individuals rather than as part of a team committed to common goals. Part of the problem has been the increasing use of specialisation in project teams. Belbin (1993), in his research on team roles at work, noted that the specialist is the 'dedicated, single-minded professional who provides knowledge and skills in rare supply'. These specialists have to learn to work together if they are to achieve successful project completion. Under traditional forms of contract, the architect will be given the management role of both the design and construction phases of the project, but will have no financial responsibility. This renders the designers not accountable for the consequences of their management, not only in terms of quality, but also in terms of cost and time. However, this is not acceptable in a partnering arrangement, where success depends on the commitment of every party to the common goals.

The public sector has been cautious about partnering arrangements because of concerns regarding government policy or European Commission directives. Recently, the UK government issued a strategy for government procurement which clarifies the ground rules. It states that partnering within the public sector will be possible if:

- ❏ it does not create an uncompetitive environment;
- ❏ it does not create a monopoly;
- ❏ the partnering arrangement is tested competitively;
- ❏ it is established on clearly defined needs and objectives over a specified period of time;
- ❏ the construction firm does not become over-dependent on the partnering arrangement;

Partnering changes mind-sets by helping all involved in the construction process to redirect their energies and focus on the real issues associated with achieving the ultimate objective. However, it is not a panacea. It is a challenging endeavour. The participants must be committed to change and to working in a team environment that fosters 'win–win' relationships.

## 16.8 Summary

Under a partnering arrangement, all parties agree from the beginning to a formal structure that will focus on creating cooperation and teamwork, in order to avoid adversarial confrontation. Working relationships are built carefully, based on mutual respect, trust and integrity. Partnering can provide the basis for a 'win–win' approach to problem solving.

Partnering depends upon an attitude adjustment, where the parties to the contract form a relationship of teamwork, cooperation and good faith performance. It requires the parties to look beyond the bounds of the contract and to develop a cooperative working relationship that will promote their common goals and objectives. The aim is not to change any contractual responsibilities, but rather to focus on what really makes the contract documents work: the relationships between the participants.

## References

AGC (1991) *Partnering: A Concept for Success*, Associated General Contractors of America, Washington, DC.

Belbin, R. (1993) *Team Roles at Work*, Butterworth–Heinemann, Oxford.

Bennet, J. and Jayes, S. (1995) *Trusting the Team: The Best Practice Guide to Partnering in Construction*, Centre for Strategic Studies in Construction, University of Reading, Reading.

CII (1995) *Dispute Prevention and Resolution Techniques in the Construction*

*Industry*, Construction Industry Institute, Dispute Prevention and Resolution Research Team, Research Summary No. 23–1, Austin, TX.

CII (1996) *Model for Partnering Excellence*, Construction Industry Institute, Partnering Task Force, Research Summary No. 102–1, Austin, TX.

ECI (1997) *Partnering in the Public Sector*, European Construction Institute, Loughborough.

Egan, Sir John (1998) *Rethinking Construction*, Report, Department of the Environment, HMSO, London.

Moore, C., Mosley, D. and Slagle, M. (1992) Partnering: guidelines for win–win project management, *Project Management Journal*, **XVIII**(1), pp. 18–21.

US Army Corps of Engineers (1991) *Construction Partnering: The Joint Pursuit of Common Goals to Enhance Engineering Quality*, US Army Corps of Engineers, Omaha, NE.

## Further reading

Baden Hellard, R. (1995) *Project Partnering: Principle and Practice*, Thomas Telford, London.

Charlett, A.J. (1996) *A Review of Partnering Arrangements Within the Construction Industry and Their Influence on Performance*, International Council for Building Research Studies and Documentation (CIB), W89 Beijing International Conference, Beijing, China
http://www.bre.polyu.edu.hk/careis/rp/cibBeijing96/papers/130_139/138/p138.htm

CIB (1997) *Partnering in the Team*, Construction Industry Board, Working Group 12, Thomas Telford, London.

CIB (1998) *Fact Sheet on Benchmarking*, Construction Industry Board, London. http://www.ciboard.org.uk/factsht.htm.

CII (1995) *Use of Incentives*, Construction Industry Institute, Implementation Status Report, 1995 CII Conference, Austin, TX.

HM Treasury (1998) Central Unit on Procurement Guidance Note 4, *Teamworking, Partnering and Incentives*, HMSO, London.

Latham, Sir Michael (1994) *Constructing the Team*, Final Report of the Government/Industry Review of Procurement and Contractual Arrangements in the UK Construction Industry, HMSO, London.

# Chapter 17

# Private Finance Initiative and Public–Private Partnerships

This chapter considers the concession contract. In this form of procurement, the contractor effectively becomes the promoter and in addition to performance, also finances the project and operates and maintains it over a sufficient period of time to generate a commercial return.

## 17.1 Concession contracts

There has been a growing trend in recent years, both in the United Kingdom and overseas, for principals, usually governments or their agencies, to place major projects into the private sector rather than the traditional domain of the public sector, by using concession or build–own–operate–transfer (BOOT) project strategies. The adoption of this form of contract strategy, often referred to as a concession contract, has led a number of organisations to consider its implementation for different types of facilities, on both a domestic and an international basis and by speculative or invited offers.

Privatised infrastructure can be traced back to the eighteenth century, when a concession contract was granted to provide drinking water to the city of Paris. During the nineteenth century, ambitious projects such as the Suez Canal and the Trans-Siberian Railway were constructed, financed and owned by private companies under concession contracts.

The transfer element of a BOOT project implies that, after a specified time, the facility is transferred to the principal and cannot be considered as real privatisation. In a BOO project, however, ownership of the facility is retained by the promoter for as long as desired, and is therefore more consistent with the concept of privatisation.

In the late 1970s and early 1980s some of the major international contracting companies and a number of developing countries began to explore the possibilities of promoting privately owned and operated infrastructure projects financed on a non-recourse basis under a concession contract.

The term BOT was introduced in the early 1980s by the Turkish Prime Minister, Turgat Ozal, to designate a 'build, own and transfer' or a 'build, operate and transfer' project. This term is often referred to as the Ozal formula.

## 17.2  Definition of concession projects

A build–own–operate–transfer (BOOT) project, sometimes referred to as a concession contract, may be defined as:

> a project based on the granting of a concession by a principal, usually a government, to a promoter, sometimes known as the concessionaire, who is responsible for the construction, financing, operation and maintenance of a facility over the period of the concession, before finally transferring, at no cost to the principal, a fully operational facility. During the concession period the promoter owns and operates the facility and collects revenues in order to repay the financing and investment costs, m.aintain and operate the facility, and make a margin of profit.

Other acronyms used to describe concession contracts include:

| | |
|---|---|
| FBOOT | finance, build, own, operate, transfer |
| BOO | build, own, operate |
| BOL | build, operate, lease |
| DBOM | design, build, operate, maintain |
| DBOT | design, build, operate, transfer |
| BOD | build, operate, deliver |
| BOOST | build, own, operate, subsidise, transfer |
| BRT | build, rent, transfer |
| BTO | build, transfer, operate |
| BOT | build, operate, transfer |

Many of these terms are alternative names for concession projects, although some denote projects which differ from the above definition in one or more particular aspects, but which have broadly adopted the main functions of the concession strategy.

## 17.3 Organisational and contractual structure

A typical concession structure, illustrating the number of organisations and contractual arrangements that may be required to realise a particular project, is shown in Figure 17.1. The key organisations and contracts include those listed below.

(1) *Principal.* The principal is responsible for granting a concession and is the ultimate owner of the facility after transfer. Principals are often governments, government agencies or regulated monopolies. The structured contract between the principal and the promoter is known as the concession agreement. This is the document which identifies and allocates the risks associated with the construction, operation maintenance, finance and revenue packages, and the terms of the concession relating to a facility. The preparation and evaluation of a BOOT project bid is based on the terms and project conditions of the structured concession agreement (SCA).

(2) *Promoter.* The promoter is the organisation which is granted the concession to build, own, operate and transfer a facility. Promoter

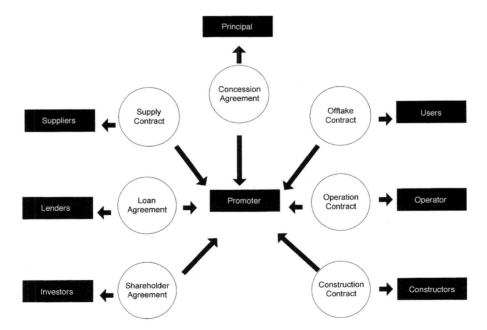

**Figure 17.1** Organizational and contractual structure for a concession contract.

organisations are often construction companies or operators, or joint venture organisations incorporating constructors, operators, suppliers, vendors, lenders and shareholders.

The following organisations and contracts may be included within the BOOT project strategy.

(1) *Supply contract*. This in a contract between the supplier and the promoter. Suppliers are often a state-owned agency, a private company or a regulated monopoly who supply raw materials to the facility during the operation period.

(2) *Offtake contract*. In contract-led projects such as power generation plants, a sales or offtake contract is often entered into between the user and the promoter. Users are the organisations or individuals purchasing the off-take or using the facility itself. In market-led projects however, such as toll roads or estuary crossings, where revenues are generated on the basis of directly payable tolls for the use of a facility, an off-take contract is usually not possible.

(3) *Loan agreement*. This is the basis of the contract between the lender and the promoter. Lenders are often commercial banks, niche banks, pension funds or export credit agencies who provide loans in the form of debt to finance a particular facility. In most cases, one lender will take the lead role in a lending consortium or a number of syndicated loans.

(4) *Operations contract*. This is the contract between the operator and the promoter. Operators are often drawn from specialist operation companies or companies created specifically for the operation and maintenance of one particular facility.

(5) *Shareholder agreement*. This is a contract between the investors and the promoter. Investors purchase equity or provide goods in kind, and form part of the corporate structure. These may include suppliers, vendors, constructors and operators, and major financial institutions, as well as private individual shareholders. Investors provide equity to finance the facility, the amount often being determined by the debt/equity ratio required by the lenders or the concession agreement.

(6) *Construction contract*. This is the contract between the constructor and the promoter. Constructors are often drawn from individual construction companies or a joint venture of specialist construction companies. Constructors can sometimes take, and have taken, the role of promoters for a number of concession projects both in the UK and overseas.

## 17.4  Concession agreements

The concession agreement is an agreement which forms the contract between the principal and the promoter, and is the document which identifies and allocates the risks associated with the construction, operation and maintenance, finance and revenue packages and the terms of the concession relating to the facility over the lifetime of the concession before transfer of the facility to the principal.

A statutory concession agreement is adopted when governments are required to ratify a treaty which may lead to legislation and consequent concessions. The terms of a concession granted under statute may usually only be altered or varied by the enactment of further legislation. Under this form of agreement, the promoter (concessionaire) would be required to enforce his rights by making an application to the courts for a judicial review of a principal's (government's) actions. In the case of concessions granted under statute, third parties may potentially have the right to apply to the courts to enforce provisions of the concession which the government do not wish to apply. A statutory concession agreement was adopted for the Channel Fixed-Link project.

The contractual concession agreement is often adopted when one government organisation enters into an agreement with a promoter to undertake a specific concession. In the contractual agreement either party may amend or relax the terms of the agreement. Under such an agreement any breach of the concession by the principal would entitle the promoter to damages, and in some cases specific performance of the terms of the concession. In the contractual concession, only the parties to the concession can enforce the terms. A contractual concession agreement was adopted for the North–South Expressway in Malaysia.

In some cases, the concession agreement is a hybrid form of both contractual and statutory elements. This form of agreement is often adopted when a principal or promoter requires an element of legislative control and the benefits associated with the contractual agreement. If, for example, planning consent is considered a major requirement for the implementation of the concession, then a statutory element may be incorporated into the hybrid form to cover this requirement.

## 17.5  Procurement of concession project strategies

Concession projects are procured by invited tender from the principal, or by a speculative bid from a promoter group to an individual principal. In the case of an invited bid, many elements of the risk may be determined by the terms of invitation. In the case of a speculative bid, the promoter will need to approach the principal to determine his obligations under the terms of the concession agreement. About 60% of concession projects are awarded as the result of speculative tendering, but within the EU this percentage is much lower.

### Speculative bids

A speculative bid is one where a promoter approaches a principal with a proposed scheme which is considered commercially viable by the promoter, and requests the principal to grant a concession to the promoter to build, own and operate a facility for a defined period of time before transferring the facility to the principal. A speculative bid is for a concession usually undertaken by the principal, and requires the promoter to prepare a concession agreement as the basis of the bid. Many concession projects begin when promoters approach governments privately to propose a much-needed project which government finds difficult to finance from the public sector budget.

In projects involving new transport infrastructure projects, a speculative bid is considered as one in which the private sector promotes an innovative project from concept to meet a perceived market need.

### Invited bids

An invited bid is one by which the principal invites a number of promoters to bid in competition for the privilege of being granted a concession to operate a facility normally undertaken by the principal or one of their organisations. An invited bid is often a solicitation of bids based on a speculative proposal made open to tender.

In the case of invited bids for transport infrastructure projects, a route will have been through a public enquiry, and the public sector seeks the support of the private sector to design, build, finance and operate this route, or where the government has defined the transport corridor and requires the private sector to select the actual route, acquire the land, and design, build, own and operate the project on the basis of a time-limited concession. In the UK, the practical consequence of privately financed projects requires each project to be authorised individually by an Act of Parliament. In invited bids such as the Dartford River crossing, a hybrid bill procedure was adopted to authorise the project's go-ahead, and the invitation involved:

❑ the identification by the government of a corridor for the proposed route;

❑ a competition for the financing, building and operation of a road serving the corridor, inviting bids from private companies;

❑ the promotion by the government of a hybrid bill to authorise the road, the tolls, land acquisition and arrangements for the concession.

Competitive bids for concession projects should follow the normal procedures of awarding public works projects. Ideally, the government would identify the project and define the project specifications, the nature of government support, the proposed method of calculating the toll or tariff, the required debt/equity ratio and other parameters for the transaction. The government would then invite preliminary proposals, with the winner being selected on the basis of normal competitive criteria such as price, experience and track record of the promoter, or on the basis of side benefits for the host country.

## 17.6 Concession periods

Concession periods are sometimes referred to as a lease from government. The concession period normally includes the construction period as well as the operation and maintenance period before transfer to the principal. However, in the case of the Shajiao Power Plant constructed in China, the concession period of 10 years excluded the construction time. In the Macau Water Supply Project, the concession period of 35 years included the refurbishment of existing plants rather than the construction of new facilities, but it also permitted the principal to repurchase the rights if deemed necessary. Such an agreement also permits the concession to be extended by mutual agreement of the parties, which in effect constitutes an addendum to the contract. A provision for an extension of the concession period, in this case the operating period, may be included in the concession agreement to protect a promoter against a principal defaulting on contractual obligations, which may result in projected returns not being met. Typical concession periods range from 10 to 55 years when granted by governments under a concession initiative.

In infrastructure projects, the concession period is often longer than in industrial facilities. Concession periods of between 10 to 15 years may be financed successfully, but principals need to accept that the project economics must be strong enough to bear the enhanced depreciation rate over such short periods, as well as the return required on the capital investment. In the Dartford crossing project, the concession period was a maximum of 20 years, which could be terminated at no cost to the Principal as soon as surplus funds had been accrued by the promoter to service all outstanding debt.

The concession period should be sufficient to:

❑ allow the promoter to recover his investment and make sufficient profit within that period to make the project worthwhile;

❑ but not allow the promoter to overcharge users when in a mono-polistic position having already recovered his investment and made sufficient profit.

The commercial viability of a concession contract, and the difficulty and uncertainty of predicting revenues over long periods of time, is a major obstacle to many promoter organisations. If there is flexibility in the concession agreement to adjust the concession period, then this may reduce the promoter's risk and allow predicted revenues to be achieved.

## 17.7 Existing facilities

In a number of concession contracts, an existing facility is included as part of the concession offered or requested. This may be a requirement of the principal in an invited bid, or offered as an incentive by a promoter in a speculative bid. In some industrial/process facilities, however, a promoter may only agree to tender a particular project provided he is given operational control of an existing facility which may effect the performance of the facility to be tendered.

The operation of an existing facility often guarantees the promoter an immediate income which may reduce loans and repay lenders and investors early in the project cycle. The commercial success or failure of an existing facility must be considered by the promoter at the bidding stage in order to determine the success or failure of the proposed concession.

Principals can influence pricing mechanisms by making existing facilities available to promoters, who are capable of earning revenues during the construction period. In the case of the Sydney Harbour Tunnel project, revenues generated by the existing bridge crossing are shared between the principal and the promoter, which enables the Promoter to generate income to service part of the debt prior to completion of the tunnel.

Assets capable of producing earnings which can be used to pay capital costs, debt service and operating expenses are a familiar feature of concession projects. A concession to operate a section of an existing highway generated US$16 million to the promoter of the North–South Expressway in Malaysia. The concession to operate existing tunnels as part of the Dartford Bridge crossing concession offered an existing cash-flow, but required the promoter to accept the existing debt on those tunnels. An existing concession also formed part of the concession agreement for the Bangkok Expressway; this arrangement required the operation and maintenance of an existing toll highway, with generated revenues being shared between the principal and the promoter.

## 17.8  Classification of concession projects

Concession projects may be classified on the basis of the method of procurement, the type of facility, the location of the facility and the method of revenue generation.

### Speculative – invited

In the case of a speculative bid, the promoter will determine which costs and risks should be borne by his own organisation, and which by the principal and other parties involved. In an invited bid, the principal will determine the concession and the costs and risks to be borne by the promoter under the terms of invitation.

### Infrastructure – industrial/process

The components of each of the packages of an infrastructure project will contain different costs and risk levels to those considered for an industrial or process plant. An infrastructure project may require large capital expenditure during construction, but operate on a small budget. A process or industrial facility, however, may require a low capital expenditure, but require a high operating budget over the operation phase.

The number and types of risks associated with a concession project may often be determined by the type of facility and the number of contracts and agreements to be included.

Infrastructure facilities may often be considered as static or dynamic facilities. A road or fixed-bridge facility may be considered as static, since the facility offers no moving parts and requires no input of raw materials or power. A static facility will usually have a smaller operation and maintenance cost than a dynamic facility. A light transit railway facility will require a major power source during the operation phase, which will result in operational costs being higher than those of a static facility.

### Domestic – international

The location of a project will determine the host country's political, legal and commercial requirements, which will be a major factor in project sanction. In the case of a domestic project, the promoter will often be aware of the country's requirements and have access to local financial markets. In international projects, promoters may need to carry out in-country surveys to determine the risks associated with meeting the requirements of the concession and how revenues may be repatriated to service loans. In effect, each international project will be determined by the constraints of the host country's government.

### Market led – contract led

One of the major risk areas associated with concession projects is the generation of revenues, which often leads to market-led revenues being far more uncertain than those based on predetermined sales contracts. The commercial risks to be considered in a concession project are often determined by the revenue classification.

The demand for a toll road which depends solely on revenues from users may be much lower than was forecast at the feasibility stage. This may be due to increased costs of fuel, or a reluctance to pay tolls by users. In the case of a water treatment plant, revenues will be contract-led, and provided demand is met and an effective price variation formula takes inflation into account, a promoter may consider the risk associated with revenue negligible compared with other risks.

In a number of market-led projects, promoter organisations will often seek contract-led revenue streams to reduce the risks associated with revenue generation. In a toll road facility, promoter organisations may approach haulage contractors and enter into take-or-pay agreements for the use of the facility. In effect, the promoter organisation is providing lending organisations with a guaranteed source of revenue, which will reduce the risks associated with the finance package.

In summary, the number of organisations, contracts, data and resources required to meet the project, and the major risk areas, will be determined by the classification of the project.

## 17.9  Projects suitable for concession strategies

Concession projects are a means of meeting the needs associated with population growth such as housing, water sanitation and transportation, with industrial growth such as power, infrastructure and fixed investments, and with tourism and recreation such as airports, hotels and resorts, and environmental concerns such as waste incineration and pollution control. Developing nations are receptive to the idea of funding such projects under a concession strategy, which will often reduce the capital and operating costs and also reduce the risks normally borne by the principal. Provided sufficient demand exists for these projects, revenue streams can be identified and the commercial viability can be determined by promoters and lenders.

The two most fundamental constraints on project development are economics and finance. In a concession project, the promoter must cover operating expenses, interest and amortization of loans, and returns on equity from project revenues. However, promoters often consider the suitability of a project based on global market forces, and the commercial viability of a project which affects the profitability rather than the facility itself.

Any public service facility which has the capacity to generate revenues through charging a tariff on throughput may be considered suitable for a concession strategy provided that suitable financing can be achieved. The most successful concession projects will be those in the small to medium range up to US$500m, as private sector equity requirements for such projects are usually obtainable.

Tolled highways, bridges and tunnels, water, gas or oil pipelines, and hydroelectric facilities are considered suitable projects, since a private economic equilibrium is obtainable. However, subsidies are often necessary for high-speed train networks, light rail trains and hectometric transport, since the prices paid by users are often low, and governments generally prefer to control prices.

The characteristics of concession projects are particularly appropriate for infrastructure development projects such as toll roads, mass-transit railways and power generation, and as such have a political dimension of public good which does not occur in other private financed projects.

## 17.10  Risks fundamental to concession projects

The two types of risk which are fundamental to concession projects are elemental risks and global risks, which are defined below.

(1) Elemental risks are those which may be controlled within the project elements of a concession project.
(2) Global risk are those outside the project elements which may not be controllable within the project elements of a concession project.

Many of the global risks are addressed and allocated through the concession agreement, with elemental risks either being retained by the promoter or allocated through the construction, operation and finance contracts.

There are two phases when risks associated with financing concession projects occur, i.e. the construction phase and the operation phase. These two distinct phases are considered as:

❏ the pre-completion phase relative to construction risks;
❏ the post-completion phase relative to operational risks, with the first few years of operation being the major operation risk.

Promoters are exposed to risks throughout the life of the project, and these can be summarised as:

❏ failure at several stages of the project;
❏ failure in the later stages of the project when considerable amounts of money have been expended in development costs;
❏ failure of the project to generate returns or the opportunity to recover costs.

Risks associated with market prices, financing, technology, revenue collection and political issues are major factors in concession projects. Risks encountered on concession projects may also include physical risks such as damage to work in progress, damage to plant and equipment, injury to third persons, and theoretical risks such as contractual obligations, delays, force majeure, revenue loss and financial guarantees.

The major risk elements of a concession project may be summarised as:

❏ completion risk, i.e. the risk that the project will not be completed on time and to budget;
❏ performance and operating risk, i.e. the risk that the project will not perform as expected;

- cash flow risk, i.e. the risk of interruptions or changes to the project cash flow;
- inflation and foreign exchange risk, i.e. the risk that inflation and foreign exchange rates will affect the project costs and revenues;
- insurable risks, i.e. risks associated with equipment and plant are commercially insurable risks;
- uninsurable risks, i.e. force majeure;
- political risk, i.e. risks associated with sovereign risk, and breach by the principal of specific undertakings provided in the concession agreement;
- commercial risk, i.e. risks associated with demand and market forces.

Demand risks associated with infrastructure projects are much greater than those for facilities producing a product off-take, since an infrastructure project is static and cannot normally find another market, whereas a product may be sold to a number of off-takers throughout the life of the concession. Facilities producing an off-take bear the risk of product obsolescence, and competition usually leads to dominant market risks, especially when operation and maintenance costs are high and concession periods are short.

## 17.11 Concession package structure

A concession project structure is a highly sophisticated structure requiring the full participation of all the parties involved in identifying and allocating the relevant project risks and responsibilities, and an appreciation of the political, legal, commercial, social and environmental considerations which have to be taken into account when preparing concession project submissions. The process of developing a concession project is immensely complicated, time-consuming and expensive.

The major components of a concession project include the agreement to:

- build, i.e. design, manage project implementation, carry out procurement, construct and finance;
- own, i.e. ownership of licence, patents, existing facilities or site;
- operate, i.e. manage and operate plant, carry out maintenance, deliver product or service, and receive off-take payments;
- transfer, i.e. hand over plant in an operating condition at the end of the concession period.

The number of components and their timing over the concession period need to be identified at an early stage of a project. This should be addressed in a format that can be utilised to identify the obligations and risks of each organisation involved in the project, so that an equitable risk allocation may be determined.

Concession contracts may be determined by the four major packages described below.

(1)  *Construction package:* containing all the components associated with building a facility, normally undertaken in the pre-completion phase, and may include feasibility studies, site investigation, design, construction, supervision, land purchase, commissioning, procurement, insurances and legal contracts.

(2)  *Operational package:* containing all the components associated with operating and, where applicable, owning the facility, and may include operation, maintenance, training, off-take, supply, transfer, consumables, insurances, guarantees, warranties, licences and power contracts.

(3)  *Financial package:* containing all the components associated with financing the building, and in some cases the early stages of operation, and may include debt finance loan, equity finance loan, standby loan agreements, shareholder agreements, currency contracts and debt service arrangements.

(4)  *Revenue package:* containing all the components associated with revenue generation, and may include demand data, toll or tariff levels, assignment of revenues, toll or tariff structures and revenues from associated developments.

This structure incorporates all the components of a concession contract into discrete packages over both the pre-completion and post-completion phases of the concession period. The type of facility, its location and revenue realisation would effectively be contained in one of the packages. Having identified and allocated the components into the four packages, a promoter organisation could then determine the risk associated with each package and how such risks would be shared. The package structure provides a rational basis for financial appraisal of concession contracts, for the allocation of risk within the concession agreement and contractually between the parties concerned, and for the structure of the tendering process.

## 17.12  Advantages and disadvantages of concession projects

A concession project may offer both direct and indirect advantages for developing countries, i.e.:

❏ promotion of private investment;
❏ completion of projects on time without cost overruns;
❏ good management and efficient operation;
❏ transfer of new and advanced technology;
❏ utilisation of foreign companies' resources;
❏ new foreign capital injections into the economy;
❏ additional financial source for priority projects;
❏ no inroads on public debt;
❏ no burden on the public budget for infrastructure development;
❏ a positive effect on the credibility of the host country.

The introduction of new technologies, project design and implementation, and management techniques are considered as advantageous to developing countries, the disadvantages, however, being host-country constraints and financial market constraints.

A major advantage of a concession project is the financial advantage to a government, as its off-balance-sheet impact does not appear as a sovereign debt. The advantage to an overseas principal is that they do not need to compete for scarce foreign exchange from the state purse, there is a specific need for the project, and risks are transferred to the promoter. The most important attractions to governments of Asian developing countries is off-balance-sheet financing, transfer of risk, speedy implementation and an acceptable face of privatisation.

The involvement of the private sector and the presence of market forces concession schemes ensure that only projects of financial value are considered. There are six main arguments for concession projects.

(1) *Additionality:* this would offer the possibility of realising a project which would otherwise not be built.
(2) *Credibility:* this would suggest that the willingness of equity investors and lenders to accept the risks would indicate that the project was commercially viable.
(3) *Efficiencies:* the promoter's control and continuing economic interest in the design, construction and operation of a project will produce significant cost efficiencies which will benefit the host country.
(4) *Benchmark:* the usefulness to the host government of using a con-

cession project as a benchmark to measure the efficiency of a similar public sector project.

(5)    *Technology transfer and training:* the continued direct involvement of the project company would promote a continuous transfer of technology, which would ultimately be passed on to the host country. A strong training programme would leave a fully trained local staff at the end of the concession period.

(6)    *Privatisation:* a concession project will have obvious appeal to a government seeking to move its local economy into the private sector.

There are three main arguments against concession projects.

(1)    *Additionality:* commercial lenders and export credit guarantee agencies will be constrained by the same host country risks whether or not the concession approach is adopted.

(2)    *Credibility:* this benefit may be lost if the host government provides too much support for a concession project, resulting in the promoter bearing no real risk.

(3)    *Complication:* a concession project is a highly complicated cost structure which requires time, money, patience and sophistication to negotiate, and bring to fruition. The overall cost to a host government is greater than that for traditional public sector projects, although proponents of the concession approach argue that overall costs are less when design and operating efficiencies are taken into account and compared with public sector alternatives.

Although there are a number of advantages and benefits associated with concession projects, very few concession proposals have reached the construction stage. A review of concession schemes by an EU Commission concluded that there were three key problems associated with concession projects, i.e. the availability of experienced developers and equity investors, the ability of governments to provide the necessary support, and the workability of corporate and financial structures.

The risks associated with concession projects are far greater than those considered under traditional forms of contract, since the revenues generated by the operational facility must be sufficient to pay for construction, operation and maintenance, and finance. The uncertainty of demand, and hence revenues, the cost of finance, the length of concession periods, the levels of tolls and tariffs, the effects of commercial, political, legal and environmental factors are only a few of the risks to be considered by promoter organisations.

## 17.13  The origins of PFI

The private finance initiative (PFI) was a policy born out of a series of privately financed projects, beginning with the Channel Tunnel project in 1987. Other transport-orientated projects followed swiftly: the second Severn Crossing, the Dartford Bridge, the Skye Bridge, the Manchester Metrolink and London City Airport.

A privately financed public sector is not a new concept. Concession contracts have been utilised in the past, with French canals and bridges being privately financed in the eighteenth century, and the railways in the United Kingdom in the nineteenth century. Toll roads were common in the eighteenth and nineteenth centuries in both the UK and the USA, where a turnpike system saw owners collect money from the users of the roads.

Many of these projects had foundered by the beginning of the twentieth century, with most major infrastructure projects having a government-financed structure. The trend back toward concessions started to emerge during the mid-twentieth century in the USA, and corresponded to a growth in the financial markets and their ability to finance such complex projects.

At present there is an interest in concession contracts throughout Europe, including toll roads in France and Spain, and power stations in Italy, Spain and Portugal. The pressure on public finance is not restricted to Europe; there is growing interest in the use of private finance all over the world, especially in the emerging economies in Asia.

## 17.14  The arguments for privately financed public services

The underlying principle behind the introduction of private sector finance has many dimensions, the obvious one being a pure private finance case where a facility and service is provided at zero, or minimum, cost to the public sector. An example of this is the second Severn Bridge, put forward by many as the perfect PFI project. It is also the public sector's exploitation of the private sector's ability to design and manage more efficiently. The public sector is characterised by substantial cost overruns and poor management skills; utilising PFI may reduce these inefficiencies.

There is a pan-European downward pressure on public capital budgets. This has forced governments to rethink their procurement strategies, with PFI-type schemes becoming more popular throughout Europe. There is particular emphasis on reducing public expenditure with regard to satisfying the criteria for monetary union under the Maastricht agreement. However, it must be noted that the targets set under Maastricht are subject to the European Union's standard measures. Therefore, reducing the public sector borrowing requirement (PSBR) through PFI will only go towards meeting this target if the PFI project '... adds value'. Hence the need to cut, or restrict, public expenditure is not just a British preoccupation, but a global objective. The rationale behind concession contracts and private finance is to achieve the following objectives.

❑ To minimise the impact of added taxation, debt burden, etc. on the finances of Governments.
❑ To introduce ... the benefits that might accrue from the use of private sector management and control techniques in the construction and operational phases of the project.
❑ To promote private and entrepreneurial initiatives in infrastructure projects.
❑ To increase the range of financial resources that might be available to fund such projects.

The reduction in public borrowing and direct expenditure are amongst the factors which encourage the adoption of concession contracts.
    Arguments against the concept include those listed below.

❑ Efficiencies may well be available in the private sector, but it is argued that such management efficiencies might not materialise, and if they do, who benefits?
❑ Construction efficiencies are also visible, with construction programmes being reduced owing to a lack of procedural constraints from Government, but the overall time from conception to operation will be far longer than for a conventional project. This is due in part to the processes involved in the evaluation and negotiations required by such projects.
❑ Financing charges are higher for the private sector than for public bodies. This reduces some of the efficiency benefits which may, or may not, occur from private sector involvement.
❑ Owing to the complexity of concession contracts, governments and their agencies have to spend large sums in advisory fees for lawyers and financiers

## 17.15 PFI in the UK

### *PFI healthcare*

There has been massive interest in the provision of services for the health sector which has attracted much negative publicity and a conspicuous absence of signed deals. According to the Private Finance Panel, there were 47 signed contracts for PFI service provision for the Department of Health by the end of 1997. These ranged from clinical waste incineration and information systems to office accommodation.

A sticking point with high-value PFI hospitals is that trusts do not appear to be very realistic in their requirements. Consortiums responding to output specifications spend large sums of money preparing bids which ultimately are not affordable. This problem arises because of the intensive level of borrowing required by the consortium at the early stages of a project and subsequent debt servicing. These costs are passed on to the trust, who find them too high.

### *PFI prisons*

The prison sector has been the beneficiary of modest PFI success with privately financed prisons. Design, construct, manage and finance (DCMF) prisons were a progression from privately operated facilities, which existed in three prisons across the UK including Her Majesty's Prison (HMP) Doncaster and Wolds Remand Prison. These prisons are still operating successfully, and showing 'significant cost benefits' over comparisons with existing public sector best performers. Alongside this was the contracting-out of prisoner escort services, which are provided by the security company Group 4. The natural next step would be to expand the number of prisons by building more. Then potential private sector operators should have an input into the design, construction and financing of the asset. The existing links within Her Majesty's Prison Service (HMPS) and experience with private operators shows that this sector is well suited to PFI treatment.

In addition to 'contractors turned operators' bidding for the project, there was the arrival of the US operator Wackenhut Correction Facilities, who joined as consortium members. Their existing track record and the privately operated prisons in the UK meant there was already in place a valuable benchmark by which to evaluate the bids. The success of the pathfinder projects was aided by the government's desire to see PFI realised, and the help they provided reflected this. In the Fazakerley and Bridgend prisons, the government agreed to act as insurers as a last resort if the operators could not obtain commercial insurance. The follow-up projects, particularly of Lowdham Grange, have hit problems due to the governments' refusal to repeat this agreement. This potentially terminates the concession agreement with premier prisons unless they can reassure the insurers and point to their record of riots and disturbances in Doncaster. It was felt by HMPS that the success of Fazakerley and Bridgend meant that they could transfer greater risks in forthcoming projects. They felt that commercial insurance could be an incentive to the operators to maintain an exemplary record when it came to discipline, as no-claims bonuses could be earned.

### Local authorities

Local authorities have been granted £250 m worth of help to cover advisors' fees to get the PFI process up and running in this sector. Local authorities have been looking at projects varying from small-scale heating projects to high-value transport interchange schemes. Along with other sectors, doubts have been expressed on the viability of smaller-scale projects. Some contractors have introduced a lower limit for bidding for PFI projects. One major contractor has set a minimum of £5 m, which could effectively prevent such local authority schemes from getting off the ground. Examples taken from other countries could show the way forward for schemes such as district heating systems. A project in Germany has seen a developer award a 10-year concession to a small contractor to finance, supply and maintain a heating system to two apartment blocks. The interesting points about this project were that the capital cost amounted to slightly in excess of DM 15 000 and the project had a net present value (NPV) of just DM 2860 over 10 years at a 10% discount rate.

## 17.16 Bidding and competition

This section provides an explanation of the EU procurement rules and how they are applicable to PFI. They are only covered in brief, and the reader should refer to Farrell (1999) for a more comprehensive understanding of the topic. These regulations are applicable to local authorities and utilities, and hence they have a significant impact on the way facilities and services are procured under PFI. The procurement strategies to which they apply are:

❑  works contracts;
❑  works concession contracts;

- subsidised works contracts;
- service contracts;
- design contests.

It should be noted that service concession contracts are defined as, 'contracts under which a purchaser engages a person to provide services and gives him the right to exploit the provision of these services to the public', and are not subject to the regulations. However, even though PFI is concerned with the procurement of services rather than assets, it still falls under the regulations applying to works projects. The reason for this inclusion is the typically high level of capital expenditure in the construction phases of projects. It is deemed that the payments for services are taking this expenditure into account, thus PFI projects are treated from the standpoint of a capital works project.

The procurement rules mean that the procedure should follow one of three routes:

- open procedure;
- restricted procedure;
- negotiated procedure.

The open procedure allows everyone who has expressed an interest in the project to bid for it. This type of approach is not suitable for PFI projects, as the cost of bidding is too high to allow mass bidding. More widespread is the use of a restricted procedure where bidders express an interest and prequalify in order to bid for a project. The last type of procedure is where the preferred bidder negotiates the terms of the contract, but usually after winning in a competitive process. The last two methods must take place in an environment where competition is satisfactory, i.e. there must be several bidders in the competition. This number should not normally be less than three in the case of a restricted procedure and five under the open procedure.

Public authorities are usually limited to using the first two methods, and can only use competition and the negotiated approach in the following circumstances:

- if pricing is not possible for the works and services;
- if specifications for services cannot be precise, so that the open or negotiated procedure could be utilised.

The preferred bidder will win the right to negotiate the contract terms, but negotiations can be conducted directly without the need for tenders to be submitted. Competition is not always necessary. In some cases a negotiated non-competition procedure is allowable if:

❑ technical or artistic reasons mean no-one else can do the job;
❑ exclusive rights exclude other bidders;
❑ it is a works concession contract;
❑ goods are manufactured for certain research, experimental, study or development purposes.

If a negotiated approach is to be used, it is important that it adheres to the following principles:

❑ transparency;
❑ objectivity;
❑ non-discrimination;
❑ equality of treatment;
❑ measurability.

In addition to the method of procuring a service, the contract duration may be subject to regulation under Article 85 of the Treaty of Rome. This addresses the concern that an excessively long contract period may distort or restrict competition. This is particularly problematic because of the length of contract periods required to secure the financing of concession-type projects.

## 17.17 Output specification

One of the guiding philosophies of the PFI is that the public bodies allow the private sector the freedom to determine how they are to provide a service to meet the relevant specification. In order to give the private sector this freedom, the principal has to be very careful in the way the requirements are expressed. This is achieved through the drafting of an output specification. This puts public bodies in an unusual situation, as they have to be the authors of a service provision document rather than an asset provision document. Although this may seem straightforward, it is sometimes difficult to separate the service from the facility, and this leads to preconceptions of how a service should be run and from what type of facility. Such presumptions should be avoided at all costs, as they remove the freedom of the private sector to innovate and produce alternative solutions.

The private finance panel (PFP) supplies the following examples:

❏ a computer system is not an output, the information service it supplies is;

❏ a particular office building is not an output, a supply of services accommodation is;

❏ a school gym is not an output, regular access to sports facilities is.

In order to define needs successfully, the approach must be carefully thought out and biased towards services provision. Clear specifications should comply with the following list according to PFP guidance:

❏ a clear statement of requirements which is concise, logical and unambiguous;

❏ a statement of what you want, not how it is to be provided;

❏ allow for innovation from tenderers;

❏ provide all the information a tenderer needs to decide and cost the services they will offer;

❏ provide equal opportunities for all tenders to offer a service which satisfies the needs of the user and which may incorporate alternative technical solutions;

❏ solutions to be evaluated against the defined criteria;

❏ an indication of the information needed in order to monitor that the service is being procured in compliance with the contract.

The issuing body must define its core requirements for the project, and it is important that these are clearly and accurately expressed. For example, a local authority may want a new school, but a consortium may develop a bid that includes other facilities such as a leisure centre and an adult education centre. Most PFI projects can be negotiated, as this allows bidders to develop and refine the specification, but there will come a point where the specification must be fixed, thus allowing bidders to develop the best way of achieving the requirements.

## 17.18  Financing public–private partnerships

Increasingly, the projects which the principals wish to realise are not sufficiently robust to be procured by total private finance funding. This gap between commercial financial analysis and social cost–benefit analysis can only be closed by some type of public sector involvement.

The public sector has four main mechanisms for participation, which operate within a strict hierarchy, as shown in Figure 17.2. Essentially, and in sympathy with the concept of encouraging the private sector to

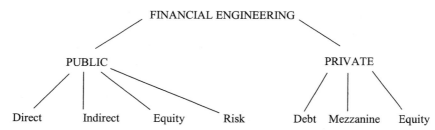

**Figure 17.2** Public–private financial engineering.

take responsibility for project risk, the public sector wishes to make the minimum possible contribution which is sufficient to reduce the investment risk to levels acceptable to the private sector.

As the first stage, the public sector can consider additional risk sharing. Risk is already shared through the concession agreement, using the basic principle that the party best able to control or manage a risk should take responsibility for that risk. For risks which cannot be controlled by either party it is usual for the public sector principal to take responsibility because the risk will only have to be paid for should it occur, rather than a premium being paid to a contractor to cover the possibility of it occurring. Therefore, by assuming responsibility for some additional risk, this may enhance the robustness of the project cash flow and hence attract investors and lenders at an acceptable rate of interest.

Once the risk-sharing option has been exhausted and further responsibility for risk is unacceptable to the public sector, the option of additional equity can be considered. Often this will take the form of offering the promoter the use, revenue, operation and maintenance of existing facilities. This stream of revenue at a critical stage of the project can be valuable in closing the gap between debt finance and the equity provision which the promoter is able to raise.

An alternative mechanism was adopted for the new Athens airport at Spata. When originally proposed as a private sector project, the debt financiers required equity funding in excess of the levels that any of the tendering consortia were prepared to offer. The Greek government then suggested a deferral of the project, and during this period a hypothecated tax on all airline tickets would be raised. This funding was then utilised as a second source of equity to bridge the gap between the level required by the lenders and the equity raised by the promoter.

Typical indirect options include tax holidays, grace periods and soft loans. A tax holiday is a predetermined period of time over which the promoter will not be liable for tax on the concession. Conventionally, each project is treated as a 'ring-fenced' investment and, in the UK, is liable to be taxed on its operating profits in the same way as any other commercial enterprise. In a number of UK infrastructure projects, a tax holiday of 5–10 years is likely to be most beneficial, and after that period the returns diminish.

Grace periods and soft loans can be used separately or jointly to assist the financing of the project. A grace period is fixed length of time before which a loan has to be repaid. Obviously the concession contract has the maximum capital lock-up towards the end of the construction phase, and if a loan can be effectively extended without interest, this can be extremely beneficial to the cash flow of the concession. As most infrastructure construction takes between 2 and 3 years, it is a grace period of between 5 and 7 years that is of most benefit. Once the infrastructure is completed, the initial debt financing, which carries high interest rates which reflect the risks associated with construction, can be repaid, and the debt can be re-financed at lower rates of interest commensurate with the known risks of operating similar infrastructure projects.

A soft loan is a loan offered at rates below the commercial market rate. The public sector loses some of the return which could have been gained, but assists the promoter by reducing the debt burden. The range of these options can be quite considerable; for example, consider the financial implications for two concession consortia, one with a public loan at commercial rate over 12 years, and another with the same loan but with a 10-year period of grace and interest at a quarter of the commercial rate.

Finally, if no other course of action is available, the public sector can provide direct financial investment. This may be necessary on projects perceived to be high risk, or projects with a marginal cash flow. One example is the M1 Vienna to Budapest Toll Road in Hungary, where the National Bank of Hungary, the Nemzeti Bank, made a contribution of 40% of the project value in Hungarian Florints. Similarly, most light rail transit schemes in the UK and in many other countries are financially non-viable without a major financial grant from the public sector, which is often between 75% and 90% of the capital cost.

## Reference

Farrell, S. (1999) *Financing European Infrastructure: Policies and Practice in Western Europe*, Macmillan, Basinstoke.

## Further reading

Merna, A. and Smith, N.J. (1996) *Projects Procured by Privately Financed Concession Contracts*, Vol. 2, 1st edn, Asia Law & Practice, Hong Kong.

Merna, A. and Owen, G. (1998) *Understanding the Private Finance Initiative*, Asia Law & Practice, Hong Kong.

Smith, N.J. and Merna, A. (1997) The private finance initiative, *Engineering, Construction and Architectural Management*, **4**(3).

# Chapter 18

# Aspects of Implementing Industrial Projects

The current demand for a higher quality of living and greater technological development has meant that large engineering projects are increasing in number and complexity year by year. The unique nature of each individual project makes successful project management a difficult task. These projects can be regarded as investments with significant risks, and organisational and engineering implications. This chapter concentrates on this type of project management, and offers general guidance for the project manager for both domestic and international projects.

## 18.1 Multi-disciplinary projects

The current world market recession has placed increasing demands upon the need to manage and deliver multi-disciplinary projects in a more effective way. This has inspired a plethora of project management approaches, some new and some revised versions of existing ideas such as programme management and concurrent engineering. The more complex the projects and the greater the number of parties involved, the greater the corresponding need for better project management, although little progress has been made.

One problem in particular is causing increasing difficulty. This concerns the abilities of large organisations to manage effectively. These large companies act in a number of roles, e.g. promoter, contractor, operator, processor, supplier and financier, and tend to be involved in a large number of concurrent projects, which are themselves composed of many sub-elements. This combination of many complex projects occurring simultaneously causes new problems for the project manager. The oil companies are prime examples of this type of organisation. This is compounded by the fact that it is difficult to identify the significant problems and risks inherent in multi-project engineering management, and to balance the cost effectiveness of corporate management, methods against new information technology developments.

It is always difficult to prejudge the evolution and development of existing systems, particularly in the more subjective management areas, but the increasing size of multi-national companies means that multi-project management is likely to become more significant. The increasing size of multi-national companies also means that this subject becomes more important, and if the trend to replace strategic company management with the project management of a large number of concurrent internal and external projects continues, then progress in this area becomes essential.

Currently, multi-project management is still regarded as an unsatisfactory balance between sophisticated project management software packages, which can view a large number of concurrent projects and optimise resources, investments and returns but without being able to incorporate the strategic company or organisational needs, and conventional corporate decision making at the highest level, which is providing long-term objective and subjective inputs to determine goals but cannot assess the implications, complexities and contradictory demands of large numbers of projects. Project management has to address the integration of these two separate levels in the traditional management structure. Flatter, more flexible organisational structures, as discussed in Chapter 8, are more suited to a project environment, but determining the priorities of tactical and strategic objectives in multi-project management remains the key issue.

The existing approaches seem to be either technically based or management based, and there is mutual distrust and misunderstanding of the relative strengths and weaknesses of the two approaches. These companies invest large amounts of money in new projects each year, and it is likely that even small savings achieved by improving multi-project management would produce large returns which could be re-invested in new projects.

Management techniques for multi-project management are the subject of research in many countries. There is a need to address the areas of the processing of information, decision analysis, inter-group communication, command structure and inter-personal relationships. The multi-project environment is unique and affects all the areas of the management system and its structure. New information is required about the structure of the multi-project environment and the effectiveness of the 'triage' concept of multi-project management decision making, which indicates the need to change the management process and priorities with time.

## 18.2 Industrial projects

The number of discrete and largely single-discipline industrial engineering projects undertaken each year is steadily increasing, and seems likely to continue to do so for the foreseeable future. Many reasons for this trend have been proposed, but there is some agreement that the most significant factor is the relatively recent combination of an increasing world population with the consequent demands for energy and raw materials, the enhanced capabilities of engineers and the availability of private sector investment finance.

For most projects a comprehensive contract is usual, based on limited competition and then negotiation of a specification, scope of work, price and other terms between the promoter and the contractor, sometimes including rights to licensed processes. International promoters usually specify their own detailed design requirements and standards, but otherwise the contractor is responsible for the design and construction to achieve the required performance, including sub-contractors' and vendors' items unless otherwise negotiated, but subject to supervision. Design is dependent upon detailed information from systems sub-contractors and vendors.

The orders for equipment to vendors in other countries mostly follow the promoters' own conditions of contract, but some use the model conditions published by the UN Economic Commission for Europe.

The usual practice is to adopt contracts with fixed-price stage payments to the main contractor, less retention. The contractor might be entitled to extra time and extra payment for a few given conditions. Many contracts combine payment on a fixed-price basis for pre-determined work, and a day-works schedule of rates per hour or per day for the use of manpower, machines, etc. when ordered.

Occasionally, entirely cost-based contracts are used, usually when the scope or nature of the construction work is undefined when the contract is entered into. For these conditions, one comprehensive contract is usual, based on a specification of the resources the contractor will have to provide and classes of work he may have to undertake, and usually written by the promoter in consultation with the prospective contractors. Instructions from the promoter then define the scope, quality and timing of the work to be done, usually progressively as decisions are made after construction has started.

Sometimes the project management and design teams of the promoter and the contractor are combined. The basis of payment to the contractor is not standardised, and is typically a combination of fixed price and reimbursable elements; monthly payment is normal, less retention.

## 18.3  Large engineering projects

It is necessary to identify the difference between a large engineering project and a complex and expensive conventional engineering project. The terms large, jumbo, super and giant are all used in project management literature, and whilst not a problem of semantics, the objective is to make clear that these projects have specialist management problems in addition to those encountered in conventional engineering management. Giant projects are so difficult to manage successfully that they should be restricted to those which cannot be realised in any other form.

An alternative approach to identifying large engineering projects is based on the concept of projects as long-term financial investments with considerable engineering implications. These investments should be optimised, which may result in the design and construction phases of the project being executed at above minimum cost in order to satisfy the project's objectives.

The management of large engineering projects has not been particularly successful; several projects have been abandoned at various stages of completion. Many of the projects that are completed break even, and only a few are profitable. Moreover, the major losses and gains do not usually occur in the operation phase of a project, but in the project start-up and commissioning. Therefore, the methods and techniques applicable to giant project management would appear to be significantly different from those needed for a standard project.

For large projects finance is important, and there should be strong credit support from the project's promoters. Although this will usually form a minor share of the total financing package, e.g. 20–30%, it demonstrates the confidence of the promoters in the project. Investors would be unwilling to consider a project which had little equity capital. Project risks, technological risks, supply of materials, debt service and force majeure are acceptable, but equity risks, creditworthiness, indirect completion guarantees and security are not.

There are immense differences between public and private large engineering projects. In the public sector, projects are selected by agreement from a priority list drawn up within a corporate national development framework. The use of appropriate technology is reviewed at the design stage, and appraisal for loan approval is made on the basis of a combined technical, institutional, economic and financial assessment. Only then are competitive bids sought from contractors. In contrast, the private sector selects projects on the criteria of individual judgement and market principles. A more thorough evaluation of the financial suitability is required as the project is not state-guaranteed, and a higher proportion of the funding may be supplied by the project promoter, with the risk of bankruptcy in the event of project failure.

The failure of a large engineering project can seldom be tolerated by the project promoters as it could induce corporate or national bankruptcy. These projects exhibit a dichotomy of risk by having to attract financial investment and avoid any prospect of failure, and yet involving large-scale, long-term construction and future operations in a volatile world market to return the no- or limited-recourse loans. Therefore risk analyses must commence at pre-feasibility stage and form a continuous and integrated part of the project planning.

The difference in public and private sector projects is again evident. Private sector projects have to meet strict criteria, and weaknesses at any point usually result in abandonment. However, whilst public sector projects often have a strong financial base, they also introduce the concept of worthwhileness. Worthwhileness can be assessed from either a cost–benefit analysis or a subjective points evaluation system, which might include political desirability and other comparative, rather than absolute, project parameters.

## 18.4  UK off-shore projects

For the large steel or concrete platform projects for the North Sea Oil and Gas fields, the most typical strategy has been separate design contracts in series for the main structure and for the topsides, separate fabrication contracts in parallel for the substructure and the topsides units, and a separate contract for the subsequent sea transport of all the above and their installation off-shore. The conditions of contract have been developed from those used for large industrial projects on-shore. For example, for the very large integral gravity platforms, the UK and Norwegian promoters have placed one comprehensive contract for the design and construction of the substructure, and separate contracts for systems, equipment, topsides module fabrication, etc. as above. For pipelines, the promoters have placed separate contracts for design, fabrication, laying and connecting.

For smaller projects, the tendency has been to continue the on-shore practice for industrial projects of a comprehensive contract, but of course differing from on-shore projects in that construction is in the contractor's and sub-contractors' yards up to load-out for installation off-shore.

Reimbursable terms of payment were common in the contracts for design, fabrication, installation, hook-up and commissioning for the first projects, but have tended to be replaced with fixed-price terms plus unit rates where appropriate, giving competition on price without the loss of flexibility to cope with uncertainty.

## 18.5  Legal systems in the EU countries

The terms of contracts adopted in EU countries vary for the obvious reasons of differences in legal systems and the extent to which liabilities and relationships in public works contracts are established by law. As would be expected, the public sector projects are subject to more national regulation than the commercial projects. There are substantial differences in legal regulations between the seven EU countries studied. Regulation is least in the UK. There also appear to be differences from country to country in the extent that national and EU law is enforced, but legal regulation is not obviously the main cause of differences in contract arrangements.

The design and construction of public engineering projects in all countries normally have to conform to standard specifications and codes, but the extent to which legislation requires the certification of design, public liability insurance, registration of engineers, contractors, craftsmen and others, or regulates tendering or the resolution of disputes varies. In no country does the law govern the important choices of deciding the number and scope of contracts. Nor does the law dictate how contracts are managed. Legal advice may be needed in preparing and managing a contract, but the primary task for promoters in any country is to define what project they want, state their priorities between quality, time and cost, assess what are the risks, foresee what may change, and then choose terms of contract (so far as is permitted by any mandatory legal rules) which are most likely to control any consequential problems.

The respective engineering industries in the EU states are directly affected in terms of design standards, procurement, the use of products and the free and fair competition for EU public works contracts.

Procurement methods vary between the member states, but there are three standard methods of tendering for construction work which are used in the majority of member states: open tendering, selective tendering and negotiation. Sometimes, particular works require specialised skills, and promoters have adopted the package deal, or management contract. Unlike the UK, many EU member states do not tender on the basis of precise quantities, which means that a great deal of emphasis is placed on the preparation and clarity of the written specification.

Terms of contracts vary between building, infrastructure and industrial projects in many EU countries, and they vary with the size of the project. One model set of conditions of contract dominate construction in one EU country, Germany. The variety of models used appears to be greatest in the UK, and so does the practice of modifying them project by project.

Harmonization is the process of removing the physical, technical and fiscal barriers by adjusting the community law, taxes, markets and procedures of the member states to conform to agreed Community standards. It is estimated that harmonization should produce a 5–6% increase in internal trade, which would be worth about US$250 000 m. In the short time between the publication of the Act and the creation of the Single European Market, from 1986 to 1992, it was not possible to complete all the necessary arrangements. Therefore, the main objective of harmonization is not to remove all national diversity, except when the diversity is perceived to be a barrier to the four freedoms of the Single European Market.

Engineering and technology projects are primarily affected by four key Directives, dealing with public works contracts, compliance, construction products and public supply contracts. However, the significant penetration of non-domestic markets requires the additional harmonization of the design techniques, insurance and liability, and professional qualifications. Ultimately, national design codes will be replaced by Eurocodes.

EU legislation already in force is intended to regulate the selection of contractors for commercial and public projects. Whether the trend towards EU harmonisation will ultimately result in the standardisation of conditions of contract is at present uncertain, but it is clear that there will be distinct national differences in contractual practice and procedure for the foreseeable future.

## 18.6 Innovation

Promoters, the government and others in the UK have become increasingly interested and active in innovations in contract arrangements, probably motivated by recurrent experiences of the costs of poor quality, late completion and contractual disputes, and the uncertainties of predicting whether these will occur. Innovations have followed the privatisation of what were public services in the UK because of the change towards commercial rather than political accountability. The harmonisation of EU practice provides a vehicle for potential innovation which could be beneficial for project managers.

Experienced private and public promoters and many contractors have stated that simpler engineering and construction contracts should be used in the UK. Conditions of contract have tended to become longer and longer, and increased rather than decreased the promoters' potential risks, because the terms in them have become impractical and the documents too complex for project managers or others to comprehend how to apply them. Additions made for one project become a precedent for the next, and further additions made contract by contract increase contractors' liabilities contrary to agreed principles for allocating risks. The simpler contracts that seem to be sufficient in Japan and other countries are cited. The main argument has been that it would be more cost-efficient for promoters and contractors collectively to agree on what are the essential general terms, and end the practice of adding 'supplementary conditions' for each project to meet what may be illusions about its special needs.

Obtaining successful results also depends upon prior attention to relationships with contractors in order to anticipate problems that might lead to conflict. The basis of non-adversarial contracts is not new, but some promoters appear to be unfamiliar with them and the corporate attitudes they need in order to be successful. Therefore, it is the promoter who must decide on the priorities, allocate the risks, propose the terms of contract, select the contractor and manage the uncertainties.

In order to utilise unfamiliar types of contract, a promoter usually requires additional managerial and contractual expertise. This requirement is often met by the appointment of an internal or external project manager. To be successful in using innovatory contracts, the project manager needs the authority to represent all interests in the promoter's organization.

## Further reading

Male, S.P. and Stocks, R.K. (1991) *Competitive Advantage in Europe*, Butterworth–Heinemann, Oxford.

Smith, N.J. and Wearne, S.H. (1993) *Construction Contract Arrangements in EU Countries*, European Construction Institute, TF003/4.

Stammers, J.R. (1992) *Civil Engineering in Europe*, McGraw-Hill, Maidenhead.

# Chapter 19

# Project Management in Developing Countries

Engineering generally and the construction industry in developing countries are sufficiently different to warrant the inclusion of a chapter in a book on project management. The range of types and size of construction companies is different, the environment in which they operate is different, the resources that are employed may be different, and the way projects are funded is different. This chapter first reviews some of the main issues that contribute to the distinctive nature of developing countries and how these affect projects.

## 19.1 What makes developing countries different?

Engineering projects, and the construction industry, in developing countries are significantly different from those in the developed industrialised world. The main differences are related to climate, population and human resources, material resources, finance and economics, and socio-cultural factors. Due recognition of these differences is a prerequisite for the successful management of projects in developing countries.

*Climate*

Many poor developing countries experience quite different climatic conditions to those in the temperate north. The type of project that is required, the most appropriate technology to be applied, and the way in which the project is managed can be influenced by these variations in climatic conditions. For example, communities living in hot climates have quite different requirements for power and water, giving rise to alternative approaches to the planning and design of the requisite infrastructure facilities. Climate will also affect the design and type of technology used: solar power may be a realistic alternative to thermal power generation, high temperatures and long hours of sunlight may indicate alternative forms of sewage treatment such as waste stabilisation ponds, and the design of buildings must be aimed at reducing glare from sunlight and ensuring that heat is kept out (rather than in).

During construction, it may be necessary to take precautions which are not required in cooler climates, such as chilling or adding crushed ice to the water used in mixing concrete, and paying particular attention to the curing of concrete. Planning and scheduling of construction work can also be affected by the climate – particularly when constructing roads, bridges and hydraulic structures in areas affected by heavy seasonal monsoons. The project manager therefore needs to be fully aware of the climatic implications from the very earliest stages of the project.

### *Population and human resources*

Of the six billion people currently living on this planet, less than one billion live in what the World Bank categorises as high-income countries, whilst some 3.5 billion live in low-income countries (over two billion of these in China and India). Nearly half of the world's population have an income of less than US$2 per day, and 1.2 billion must survive on less than US$1 per day. Up to 25% of children born in many low-income countries do not reach the age of five, and the average life expectancy in the poorest countries is less than 40 years. Over a billion people do not have access to a safe water supply, a third of the world's population have inadequate sanitation facilities, and nearly five million deaths per year in developing countries are directly attributable to water-borne diseases and polluted air. In addition to this lack of basic water and sanitation facilities, many live in sub-standard housing, transportation and communication links are poor or non-existent, and fuel for heating and cooking is in short supply.

The problem is compounded by exponential population growth and rapid migration from rural to urban areas worldwide. It is estimated that the world population will rise to 8.5 billion by 2025, of which over 7 billion will be in developing countries. By 2050, the total world population could increase to as much as 10 billion. In the past, much of the developing world's population has been in rural areas, but this is now changing rapidly, and population in urban areas is growing much faster than in the rural areas. In 1994, about one billion people in the developing world lived in towns and cities; it is estimated that by 2025, this will have risen to around 4 billion.

These facts and figures point to a huge need for infrastructure development throughout the developing world – a need that can only be fulfilled through the implementation of well-managed engineering projects.

Population and human resources not only affect the need for projects, but also the way projects are implemented. The large pool of available and relatively cheap labour, much of which is unemployed or under-employed, points to a less mechanised approach to construction and a greater use of human labour. Labour-intensive construction requires a different approach to the planning, design and management of projects, and these issues must be addressed at the earliest stages of the project. Although the labour force in developing countries may be plentiful, it is also likely to be relatively unskilled. The questions of training and technology transfer therefore need to be taken into consideration throughout the planning and implementation of the project.

## Materials, equipment and plant

Many of the materials commonly used in construction projects are often not readily available in developing countries. Cement and steel may have to be imported and paid for with scarce foreign exchange. Delays in importation and difficulties in gaining passage for imported goods through customs are not uncommon and need to be allowed for. Even if materials are manufactured in the country of the project, supplies cannot always be guaranteed and the quality may be inferior to that normally expected in industrialised countries. Production capacity and quality should therefore be assessed before the detailed design is done.

In some cases, it may be necessary to consider alternative materials such as stabilised soil, ferrocement, round-pole timber, or pozzolana as a cement replacement. Many such alternatives are traditional indigenous building materials and may be more acceptable than steel or concrete. An assessment of what is available and appropriate needs to be made at an early stage and used in the design.

Imported mechanical equipment, whether it be for construction or for incorporation within the completed project, is expensive and requires maintenance. Trained maintenance technicians and a reliable supply of spare parts are the absolute minimum requirements if the equipment is to continue to function over its anticipated life. There is generally a shortage of reliable and operable construction plant in developing countries, and it is often not possible to hire plant because plant hire companies do not exist. Managers of projects, particularly large ones, will have to import plant for the project, and then decide on whether to sell it or transport it back to the home country on completion of the project.

### Finance and economics

Although there is a great need for new projects in developing countries, there is also a lack of funds from the normal sources expected in the developed countries. Many projects are funded externally from national aid agencies, international development banks, or non-governmental organisations (NGOs) such as international charities. Project managers involved in the identification, preparation and appraisal stages of a funded project need to be fully aware of the requirements of the grant- or loan-awarding agency to whom they are making application for funding, as each has its own specific requirements.

### Socio-cultural factors

The successful management of a project in a developing country requires an understanding of the ways in which society is organised, and the indigenous cultural and religious traditions. In Muslim countries, time must be allowed for workers to participate in daily prayers, and during the month of Ramadan fasting is mandatory during daylight hours, thus affecting productivity and the way work is organised. The respective roles of men, women, religious and community leaders, and land owners must be understood, particularly when managing projects in which the community is actively participating.

If socio-cultural factors are not taken into consideration, a project may not be successful even if it is successfully constructed. A new water supply may not be used if the community feel they do not own it, or if traditional existing sources of water have a strong cultural significance; sanitation facilities might be under-used or neglected if the orientation offends religious beliefs, or if men's and women's toilet blocks are sited too close together. As with other factors already mentioned, a knowledge of socio-cultural influences is therefore necessary at the earliest stages of a project because they may have a significant effect on project identification, appraisal and design, as well as construction and operation.

### Working with other professionals

Project managers often have to work with a variety of professionals, e.g. electrical and mechanical engineers, chemical engineers, heating and ventilating engineers, environmental scientists, architects, quantity surveyors and planners, but nowhere is the diversity of professionals so great as when the project is located in a developing country. In addition to the list given above, the project manager working in a developing country may well have to work closely with agricultural scientists, health and community workers, educationalists and professional trainers, economists, community leaders, sociologists, ecologists, epidemiologists, local and national politicians, and perhaps many others. Good written and oral communication skills and an ability to understand the views and perspectives of other professions are valuable qualities in any project manager, but they could be vital when the project is set in a developing country.

## 19.2 The construction industry in developing countries

Unlike the developed countries, many developing countries do not have a mature construction industry consisting of well-established contracting and consulting companies. Much, if not most, of the building and construction is done by the informal sector. This consists of individual builders and tradesmen, who are mainly concerned with building family shelters, and community and self-help groups, who may construct small irrigation works, community buildings, grain storage facilities, water wells and the like. These individuals and small groups rarely attract funding, and the works are completed using labour-intensive methods and locally available materials. Small community groups may gain the attention of national and international non-governmental organisations, who are more inclined to fund and work with such groups than are formal aid agencies and development banks.

The formal sector consists of public or state-owned organisations and private domestic contractors. The proportion of work carried out by the public sector is usually much higher in the less developed countries than it is in the richer countries, but low salaries, lack of incentives and poor promotion prospects often result in highly demotivated professional and technical personnel.

Contracting is a risky business in any country, but in many poor developing countries the lack of access to financing, excessively complex contract documents, failure to ensure fair procurement practices, the high cost of importing equipment, and the fluctuations of demand for construction often mean that the private sector of the construction industry has not had the opportunity to establish itself sufficiently to bid for major infrastructure projects. These are almost entirely carried out by international contractors, and funded by national and international loans and grants. In some countries, up to 80% of major building and civil engineering is executed in this way.

Because of the importance of the construction industry to the development of the overall economy, the growth of an indigenous construction industry needs to be encouraged. In an attempt to do this, the World Bank and other agencies have encouraged the 'slicing' and 'packaging' of contracts, i.e. breaking up large contracts into smaller ones that local contractors would have the resources and capability to bid for. Other initiatives have included a greater use of labour-intensive construction, help with finance for smaller construction companies, the encouragement of plant hire companies, and management training for the principals and staff of construction companies. Technology transfer and training, particularly management training, play an important role in this development.

## 19.3 Finance and funding

Developing countries are, by definition, poor. Funding for projects will therefore be scarce, loan finance difficult to obtain, and resources scarce. In many cases the only source of finance will be from development banks, aid agencies or charitable non-governmental agencies, many of whom obtain at least part of their funding from national aid agencies.

The major development banks include the World Bank, the Asian Development Bank (ADB), the African Development Bank (AfDB) and the European Bank for Reconstruction and Development (EBRD). These are all multi-lateral funding agencies, drawing their funds from several different countries. They operate as commercial banks, loaning money at agreed rates of interest. The loans have to be repaid, albeit the loan conditions are often more favourable than those obtained from commercial banks, and they may allow a period of grace before repayments commence.

Most industrialised countries have their own government bi-lateral aid agencies, such as the UK's Department for International Development (DfID). These agencies fund projects in developing countries through loans and grants, and also direct some of their allocated funds to those development banks of which they are members. Aid awarded directly by these agencies is often 'tied', i.e. the grant or loan is conditional upon some of the goods and services needed for the project being procured from the donor country.

Loan finance for construction companies to expand, buy equipment, or simply to maintain adequate balances and ease cash flow difficulties is extremely difficult to obtain from commercial banks in developing countries. Contractors may therefore be forced to borrow from other sources at inflated rates of interest. However, some countries have development finance companies, which act as intermediaries to channel funding from external agencies such as the World Bank to the construction industry and other developers.

## 19.4  Appropriate technology

The distinctive nature of the construction industry in developing countries suggests alternative approaches to the design, construction and management of projects. The application of appropriate technology is one approach that has been promoted as a way to overcome some of the problems associated with the implementation and long-term sustainability of development projects in the Third World. Appropriate technology should be able to satisfy the requirements for fitness for purpose in the particular environment in which it is to be used. It should also be maintainable using local resources, and it should be affordable. Many would argue that all technology should be appropriate, and perhaps therefore it is intermediate technology that we should be focusing our attention upon.

The concept of intermediate technology was first developed by E.F. Shumacher (1973), who defined it in terms of the equipment cost per workplace. He suggested that the traditional indigenous technology of the Third World could be represented as a £1 technology, whilst that of the industrialised world was a £1000 technology. An example from the agricultural sector is the traditional hand or garden hoe as a £1 technology, compared with a modern tractor and plough as the £1000 technology. He pointed out that throughout the world, the equipment cost of a workplace was approximately equal to the average annual income, and that any budding entrepreneur could save sufficient money over a 10–15 year period to purchase the equipment necessary to start a small business. However, if the budding entrepreneur is a resident of a developing country, where salaries are a fraction of those in the industrialised world, it would take them over 100 years to purchase equipment of the advanced £1000 technology type, and this is clearly impossible. Schumacher therefore advocated an intermediate technology – a £100 technology (the animal-drawn plough) – which would be affordable to people in the Third World, but which would improve efficiency, reduce drudgery, and help develop and improve the economy.

In the context of construction technology, concrete can be mixed slowly and inefficiently by hand, using a flat wooden board and a shovel (£1 technology). Alternatively, a mechanised concrete batching and mixing plant can be used (£1000 technology). The labour-intensive method is very slow and the quality of the concrete is likely to be inferior. The mechanical mixer is not only expensive, but it may be difficult to maintain owing to the lack of local skilled mechanics and the difficulty and cost of obtaining spare parts. Supplies of electrical power or fuel for the mixer may also be unreliable. An intermediate technology concrete mixer, developed in Ghana, is illustrated in Figure 19.1. It consists of a box lined with thin galvanised metal and fitted with a hinged top, mounted on simple wooden wheels, and fitted with a handle. Sand, aggregate, cement and water are placed in the box, the lid is fastened and the device is then simply pushed around the site until the concrete is mixed. The device is cheap, simple to maintain and repair, requires no source of power apart from human labour, and produces quite good quality concrete – laboratory tests indicated cube strengths of approximately 90% of those obtained from a mechanical mixer.

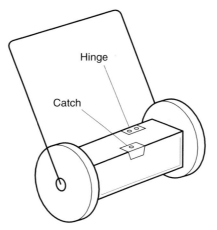

**Figure 19.1** Intermediate technology concrete mixer.

There are critics of the use of intermediate technology, including many from developing countries themselves, who point to the rapidly expanding economies of Korea, Malaysia and Taiwan as examples, and argue that without access to the most advanced technology available, the Third World will never catch up with the industrialised world. On the other hand, there are numerous examples illustrating the unsustainability of advanced technology – the factories operating at a fraction of their design capacity because of inadequate distribution networks, or lack of maintenance staff or spare parts, the water and wastewater treatment plants which are not functioning because of lack of funds for consumables and spare parts; and the impassable sealed roads which have not been resurfaced owing to the lack of suitable plant or materials.

The project manager must decide on what is or is not appropriate in any given context. They need to address the questions of fitness of purpose, maintainability, cost and sustainability. Whether 'local indigenous', 'intermediate' or 'advanced' technology is the most appropriate will depend on the physical, social, cultural and economic environment of the particular developing country in which the project is set.

## 19.5 Labour-intensive construction

One facet of appropriate construction technology is the use of less plant and equipment, and more labour. This is termed *labour-based* or *labour-intensive construction*. Much of the infrastructure, e.g. sewers, aqueducts, canals, railways, in industrialised countries was built by our ancestors using labour-based methods, and it is only fairly recently that extensive use has been made of heavy machinery for construction and building. With the change to machine-based construction, engineers and project managers, in both industrialised and developing countries, became less familiar with labour-based methods. Training for civil engineers has become based on equipment-intensive methods, and this has become the norm throughout most of the world. This situation was encouraged in many developing countries through the provision of aid, particularly bilateral tied aid, by funding agencies expecting or requiring trade and business as a condition for loans and grants.

In the 1970s it became clear that the plight of the world's rural poor was not improving, partly because the small and isolated infrastructure projects required for these rural areas were not attractive to either funding agencies or large equipment-based contractors. Research into the potential for labour-based construction revealed that there was a reluctance to adopt such methods because it was felt that the costs could not be predicted accurately, the labour force was unreliable, and it would be more expensive and more prone to delays than equipment-based construction. In addition to this were the many problems associated with the welfare of large numbers of labourers.

However, the reliance on plant-intensive construction has a number of drawbacks for developing countries, the main one being the high expenditure of foreign exchange, something that few Third World countries can afford. In contrast, labour-based construction actually generates employment and produces an income for those engaged on the project.

It has now been demonstrated that, in certain circumstances, labour-intensive methods can compete with plant-based methods in terms of both technical quality and cost. This can be made possible with good management. Good management of labour-based construction entails ensuring that the morale and motivation of workers remains high at all times by offering incentive payments, providing good training, attending to the welfare of the workers, and ensuring that materials and tools are always available and in good condition. Because labour-based construction projects can often entail work being carried out at a number of small, dispersed and remote sites, good communications between the various sites and good planning and coordination are essential.

Most labour-based construction projects will in fact use a combination of various forms of motive power – human, animal and machine. The key to efficient construction is selecting the optimum combination for a particular project. Long-distance haulage is probably best done by truck, animals are better and cheaper for shorter distances over steep or rough terrain, and even shorter haulage distances might be most suited to labourers with wheelbarrows or headbaskets. In many construction operations, a mix of labour and machine will provide the optimum solution – earthworks may use labour for excavation and a truck for haulage, for example. Balancing these resources to ensure full productivity is one of the tasks of the manager.

Some activities are always more suited to machine-based construction, whilst others are better carried out by labour-based methods. For example, although a concrete pipe culvert might be installed using machines, brick arch culverts can only be constructed using labour. An asphalt road surface requires equipment, and it would be totally impractical to consider labour-based methods for this form of construction. Earthworks, excavation and quarrying are all well suited to labour-based methods.

The implicit choice of equipment-based construction methods has an influence on the planning and design of the works, and a fair comparison between the financial costs of labour-based and equipment-based methods may be difficult once the design is finalised. Therefore, if labour-based methods are to be given an unprejudiced assessment, alternative designs suited to labour-based construction should be considered. This is sometimes referred to as *design equivalence*. For an even more comprehensive comparison, a full economic appraisal should be carried out, whereby the economic benefits of enhanced employment and the potential for growth and development are included as benefits of the project.

Finally, it is important to note that the availability of labour in rural agricultural areas can vary considerably with the seasons: during planting and harvesting time, full employment and higher wages can be obtained on the farms. It may be necessary to raise project wages at these times in order to ensure continuity of work, although this may have the disadvantage of workers refusing to return to normal rates of pay later. It may also have the added drawback to the community of decreased agricultural output. In some cases, it may be necessary to employ migrant workers from another area, and this will entail providing suitable accommodation, normally in camps, which must be provided with water, sanitation and other essential services. In any case, there is a need to collect data on the availability of labour prior to making the decision on whether or not to adopt a labour-intensive approach to the project.

## 19.6 Community participation

In industrialised countries, the general public may be involved in the sanction and approval stage of projects through public enquiries or public protest, but they are unlikely to be involved in the planning and design of projects, and it is even less conceivable that they would participate in their construction or operation. For projects in developing countries, particularly small projects in rural areas, the concept of 'community' or 'beneficiary' participation is now accepted as being expedient, if not essential, for success.

The participation of the community for whom the project was intended was initially employed to provide assistance with the construction of rural projects, such as the development of water, sanitation and irrigation schemes. This was primarily to reduce costs and ensure that the local community would have sufficient expertise to enable them to operate and maintain the schemes once they had been constructed. The idea has since been developed further to encompass the identification, planning and even design of projects, the argument being that the community have a much better local knowledge, they know what they want, and of course they are aware of all the social, cultural and religious factors which may affect the design.

For project managers this can entail a great deal of additional work, and it can delay the start and completion of a project. The reported advantages are that it will help to ensure that the project is used, and that it can, and will, be successfully operated and maintained.

In some cases, such as the improvement of water quality, rural communities may not be aware of the advantages of a project, and community training and education may be necessary. The construction of pilot schemes can be necessary in order to convince villagers of the benefits to be derived from a project and to obtain their views on the design and possible improvements. This all places an extra burden on the project manager, and highlights the importance of communication skills and the ability to work with a variety of different people and organisations.

## 19.7  Technology transfer

The Third World needs new construction technologies and management expertise in order to develop, but only if those technologies and techniques are appropriate. The long-term benefits acquired when new technology is introduced and used for one short contract will be negligible, and may possibly even be detrimental to the achievement of long-term goals. However, if there is a real commitment to the ideal of technology transfer in terms of the acquisition of knowledge, skills and equipment, the benefits may be considerable.

Effective technology transfer is more than just education and training. Although training is essential, a much greater understanding of techniques, processes and machinery will be acquired from their actual use than from merely observing or learning about their use. Even more control and mastery will be obtained by owning and being responsible for them. This can be achieved through direct purchase of equipment, entering into a joint venture, becoming a licensee or franchisee for an established process, or entering into some other form of contractual arrangement with an experienced company or organisation in the developed world.

Direct purchase of equipment involves the least amount of risk on the part of the vendor and the greatest risk for the purchaser. However, provided that good training, reliable after-sales support, established channels for the supply of spare parts, and sound warranties are included in the package, this can be a quick and effective method of transferring some technologies.

Becoming a licensee may be seen as purchasing intellectual knowledge in addition to purchasing or leasing equipment. It is usually in the interests of the licensor to provide a higher level of support than might be expected from a vendor because the terms of the licence would normally entail the payment of a percentage of the value of the work carried out. Hence some of the risk of such a venture will be taken by the licensor.

Joint ventures between an established contractor in an industrialised country and a contractor in a developing country can be a very effective way to transfer technology and encourage the development of the indigenous construction company. There are a variety of different forms of joint venture, all of which require some investment from both the partner in the developed country and the one in the developing country. The investment from the developed country partner is in the form of either cash or equipment, together with technical and management expertise. The developed country partner may invest either cash and/or premises, labour and locally available equipment. Joint ventures will only be an effective way of transferring technology if both parties are firmly committed to the idea. Very often, joint ventures are formed merely to satisfy a local requirement for overseas contractors to enter into a joint venture with a local company in order to bid for work in that country. In such cases there is often little or no commitment on either side, the overseas company being involved out of necessity, and the local company allowing their foreign partner to make the decisions while they take a back seat – and their share of the profits! Joint ventures set up for short one-off projects are likely to be a less effective means of technology transfer than long-term ventures.

Contractors working in developing countries can help the process by providing training to local staff and sub-contractors, and most funding agencies allow for training within the loans or grants awarded for projects. Agencies also provide finance specifically for training and technical assistance, which can be provided through an arrangement known as *twinning*. Twinning is a formal professional relationship between an organisation in a developing country and a similar, but more mature and experienced, organisation in another country. Unlike training programmes, these are often long-term arrangements and may involve lengthy visits from key personnel in both organisations to their counterparts in the twinned institution.

## 19.8 Corruption

Although corruption does occur in industrialised countries, it would appear to be more widespread in developing countries, and any project manager working in such areas needs to be aware of the problems and of their own professional codes of conduct and ethics. Corruption can affect a project at a number of stages in the project cycle, such as the marketing of construction services, obtaining planning permissions and building consents, pre-qualification, tender evaluation and award, construction supervision and quality control, getting materials and equipment through customs, issuing payment certificates, and making decisions on claims and completion. It is not unusual to read of buildings and other structures collapsing as a consequence of poor-quality construction and building-code infringements which have been traced to bribery and corruption.

It takes two parties to make bribery effective – the party that requests the payment and the party that provides the payment. In this regard, international companies have been as culpable as indigenous government officials in some cases. It has not been unusual for international construction companies to employ local 'agents' in order to facilitate business dealings in a foreign country. Such agents may have a perfectly legitimate function, but payments or commissions to agents can also be an effective way of concealing illicit payments to corrupt officials in order to secure a contract or to facilitate completion of the work. The World Bank has suggested that an average of around 10% of contract costs are siphoned off in bribes and other illicit payments, with up to 80% being lost in this way in extreme cases.

Until fairly recently a blind eye was turned to this problem, and some economists and business people argued that such practices actually stimulated economies by allowing more wealth to enter and be distributed throughout the country. It was further argued that bribes could cut through overly bureaucratic red tape and allow construction projects to meet important deadlines, thus saving costs in the long term.

These arguments have since been thoroughly discredited, and major international institutions such as the World Bank, Organisation for Economic Cooperation and Development (OECD), United Nations and the International Federation of Consulting Engineers (FIDIC) have all taken a firm stand on the issue. The World Bank points out that apart from money finding its way into the bank accounts of corrupt individuals and thus making less available for other development projects, bribery often means that the wrong projects are implemented, leaving a graveyard of white-elephant infrastructure facilities that may never be used to their full potential.

FIDIC is quite clear on the issue, and modified their code of ethics (FIDIC, 1996) by adding the following text.

The consulting engineer shall:

❏ neither offer nor accept remuneration of any kind which in perception or in effect either (a) seeks to influence the process of selection or compensation of consulting engineers and/or their clients, or (b) seeks to affect the consulting engineer's impartial judgement;

❏ cooperate fully with any legitimately constituted investigative body which makes inquiry into the administration of any contract for services or construction.

In their policy statement on corruption, FIDIC make it clear that the term 'bribe' includes indirect payments, using mechanisms such as scholarships, currency exchange facilities and the actions of agents, as well as direct cash payments. Not only FIDIC, but many other powerful international organisations are now prepared to act firmly by sanctioning their members, terminating contracts or taking other disciplinary action against any individual or firm that is found to be involved in corrupt practices.

In 1993, Transparency International initiated a major crusade against corruption, and further details of their work can be found at the web site listed below.

## 19.9 Summary

The world's population is still growing at an alarming rate, and much of this growth is in the less developed world. The world urban population is growing even faster, and once again it is in the developing world where the largest and fastest-growing cities are to be found. Although infrastructure development is taking place, it is barely keeping pace with the ever-expanding population, all of whom require the basic necessities of food, shelter, water and education. The need is overwhelming. Without project managers who can effectively manage engineering projects, this need will never be met. However, successful projects in developing countries require managers who recognise the needs and are knowledgeable of the clear and distinctive differences, difficulties, peculiarities and rewards of managing projects in these countries.

## Reference

Schumacher, E.F. (1973) *Small is Beautiful: Economics as if People Mattered*, Blond and Briggs, London.

## Further reading

Cooper, L. (1984) *The Twinning of Institutions – Its Use as a Technical Assistance Delivery System*, World Bank Technical Paper 23.
Côté-Freeman, S. (1999) *False Economies*, Developments, 4th Quarter, Department for International Development.

Coukis, B. and World Bank staff (1983) *Labor-Based Construction Programs*, Oxford University Press, Oxford.

FIDIC (1996) *Code of Ethics*, http://www.fidic.org/about/ethics.asp

Institution of Civil Engineers (1981) *Appropriate Technology in Civil Engineering*, Thomas Telford, London.

Smillie, I. (1991) *Mastering the Machine – Poverty, Aid, and Technology*, Intermediate Technology Publications, London.

Transparency International, http://www.transparency.org/

USAID, *Anticorruption Web Site*, http://www.usaid.gov/democracy/anticorruption/

World Bank (1984) *The Construction Industry – Issues and Strategies in Developing Countries*, World Bank.

World Bank (1992) *World Development Report 1992 – Development and the Environment*, Oxford University Press, Oxford.

World Bank (1994) *World Development Report 1994 – Infrastructure for Development*, Oxford University Press, Oxford.

World Bank, *Data and Maps*, http://www.worldbank.org/data/

World Resources Institute (1994) *World Resources 1994–95*, Oxford University Press, Oxford.

# Chapter 20
# The Future for Engineering Project Management

The discipline of project management continues to evolve over time. The recognition of project management as a profession has spread across all business sectors, from construction to IT, from time based to performance and quality based, from discrete operations to managing by projects.

In the last few years, the changes in project management have been reflected in increased documentation, manuals of good practice and minimum qualification standards. Notable amongst these are BS6079 and the Association for Project Management's body of knowledge and the Project Management Institute's (revised) body of knowledge, the International Project Management Association's agreed certification criteria for qualified project managers, and standard contracts for the engagement of a project manager in the US.

Industry is also moving towards a more collaborative form of procurement and implementation. In addition to the established use of joint ventures, consortia and management contracts, new opportunities such as the private finance initiative, partnering and alliancing have been adopted, placing duties and obligations onto the project manager. To fulfil these requirements new skills and knowledge are necessary, and it is this information which accounts for much of the new material in this edition.

The essential function of the project manager remains unchanged. The project manager should be appointed early in the process, ideally with overall control from inception through to operation or decommissioning, and should pay due attention to the role of the parties in a project and the basic guidelines.

## 20.1 The role of the parties

### The promoter

The ultimate responsibility for the management of a project lies squarely with the promoter, and consequently any project organisation structure should ensure that this ultimate responsibility can be effected. This is not to say that the promoter must be involved in the detailed project management, but it does mean that the machinery must be in place for the promoter to make critical decisions affecting the investment promptly whenever necessary.

The project manager must ensure that the promoter organisation supports the project team with direction, decision and drive, and that it regularly reviews both objectives and performance.

### The project manager

The role of the project manager is to control the evolution and execution of the project on behalf of the promoter. This role will require a degree of executive authority in order to coordinate activities effectively and take responsibility for progress. It will be necessary to define the extent of such delegated authority, and the means by which instructions will be received with regard to those decisions which the project manager is empowered to make.

Ideally, the project manager should be involved in the determination of the project objectives and subsequently in the evaluation of the contract strategy. The project manager must therefore drive the project forward and think ahead, delegate routine functions and concentrate on problem areas.

If the project manager is to fulfil the task of control of the realisation of the project on behalf of the promoter, decisions taken on engineering matters cannot be divorced from all other factors affecting the investment. Control may only be achieved by regular reappraisal of the project as a whole, so that the current situation in the design office, on fabrication, on the supply of materials, and on site may be related to the latest market predictions. If this is done, the advantage to be gained, say, from early access to the site may be equated with any additional costs in full knowledge of the value of early or timely completion. The continual updating of a simple 'time and money' model of the project originally compiled for appraisal will greatly facilitate effective control during the engineering phase of the project.

## 20.2  Guidelines for project management

The project management process is summarised briefly below.

(1)   The success of a project or contract depends on the management effort expended by the promoter prior to sanction, and by all parties prior to the award of a contract.

(2)   The promoter commits to investment in the project on the basis of the appraisal completed prior to sanction. The appraisal must be realistic and identify all risks, uncertainties, potential problem areas and opportunities. Single-figure estimates are misleading, and should be supported by figures showing the range of the likely outcome of the investment.

The overriding conclusion drawn from research is that promoters and all parties involved in projects and contracts benefit greatly from a reduction in uncertainty prior to their financial commitment. Money spent early buys more than money spent late. Willingness to invest in anticipating risk is a test of the promoter's wish for a successful project.

(3)   It is essential that project management ensures that the promoter clearly defines the project objectives, together with the ranking of their relative importance. The likelihood of a successful project is greatly improved when all key managers of design, construction and supporting groups are fully informed and committed to these objectives. The project's objectives should also be communicated to the other parties involved in project implementation. The dominant considerations must be fitness for purpose of the completed project, and safety during both the implementation and the operation phases.

Thereafter, the normal primary objectives are concerned with cost, time and quality. These are interrelated and may conflict. The fact that the promoter usually does not see any return on his investment until the project is commissioned suggests that timely completion should be a priority.

(4)   Engineering projects are normally of short duration and are completed against a demanding time scale. Adequate staff of the right quality must therefore be appointed and given training in the appropriate techniques and procedures. All staff concerned must be familiar with the contractual procedures employed. For the promoter, one person, the project manager, must be ultimately responsible, and be known to be responsible, for the realisation of the project.

(5)    Although the scope of the project will be agreed at sanction, it is probable that the conceptual design, which will determine the final layout and size of the functional units, will follow early in the engineering phase. If the conceptual design is rigorously reviewed, this provides an opportunity both for cost saving and for ensuring that the proposals meet the promoter's objectives. Particular attention should be given at this stage to the subsequent operation and maintenance of the project. Regular review or project audit of both objectives and achievement should be linked with up-dating the project plan.

(6)    Effective control of the project will only be achieved through continual planning and replanning. Management effort should be concentrated on the present and the future; time devoted to the reporting and collection of historical data should be kept to a minimum.

   In planning, the project manager must take a broad view of the project and aim to coordinate design, implementation, commissioning, and subsequent operation and maintenance. The interaction of external factors such as contractors, access, statutory requirements and public relations must all be considered.

   A contractor will plan in detail and aim to achieve continuous and efficient deployment of resources. Owing to the likelihood of change, the contractor's programme should be flexible and subject to constant review by the project manager.

(7)    A strategic project plan should clearly show the financial consequences of alternative courses of action and of indecision. It is therefore convenient to develop the plan as a time-and-money model of the project which will react realistically to changes in timing, method, content and cost of work. This realism is largely dependent on the correct definition and allocation of costs and revenues as either fixed, time-related or quantity proportional charges. Time-related costs are significant in all types of construction work and predominate in many civil engineering projects. Therefore, adherence to the programmed time schedule for the work will also control both cost and investment.

(8)    Time lost at the beginning of a project can rarely be recovered; particular attention must therefore be given to the start-up of the project. Similarly, sufficient time must be allowed for mobilisation by each contractor.

(9)    Consideration of alternative contract strategies will frequently focus attention on any deficiency of information, and on the problems which will hinder the achievement of the project

objectives. Selection of an appropriate contract strategy at an early stage of project implementation is perhaps the most important single activity of the project management team.

(10) The appointment of a contractor on the sole criterion of lowest bid price will not necessarily lead to a harmonious contractual relationship. The lowest tender may not ultimately produce the lowest contract price. Parties are making their commitment at this point, and should be fully aware of both the promoter's objectives and the contractual responsibilities. Selective tendering followed by rigorous bid appraisal, including a study of the contractor's programme and resource allocation, will do much to ensure that the contractor has not misjudged the job and that the price is realistic. The production of an operational cost estimate will greatly aid the project manager in this appraisal.

(11) Throughout the implementation period of the project, the promoter will inspect and approve the quality of workmanship of contractors and manufacturers. Again, an adequate number of staff with relevant experience must be employed. Prior definition and agreement of acceptable standards is essential, and all parties should be aware of tolerances. There is a tendency for design engineers to specify unnecessarily high standards, the achievement of which may prove difficult and/or expensive. The desired quality of workmanship must always be considered in relation to the promoter's other prime objectives, usually timely completion and economical cost.

(12) Promoters frequently underestimate both the extent and consequence of change. The project manager should rigorously assess the cost and benefit of all design changes before they are implemented. Priority should be given to timely completion of the project.

The better organised the contractor, the more likely it is that they are working to a tight, well-resourced programme. The disruptive effect of variation may therefore be serious. Modifications to manufacturing plant are sometimes best implemented during some future shut-down of the plant for maintenance, rather than by variations in the original contract.

(13) Involvement in prolonged bargaining over claims is a sign of failure. Evaluate and agree payment for variations and claims as the job progresses. The valuation should be based on prices, resource output and efficiencies similar to those incorporated in the contractor's tender.

(14) Quality assurance systems can assist in the setting and achieve-

ment of project objectives. Properly practised, the system requires precision of communication and can be considered as its greatest value to project management. Care must be taken to ensure that the adoption of a quality assurance system does not result in rigid adherence to unnecessarily demanding specifications. Neither must the system inhibit the flexibility and judgement required for the management of the uncertainties associated with the project.

(15)   In the context of project management, the quality of the performance of the project is greatly dependent on the quality of project staff. Projects are managed by people who are continuously directing and communicating with other people. Great attention must be paid to the selection and motivation of staff. Personality and an ability to think ahead are as important as technical know-how.

Management is concerned with the setting and achievement of realistic objectives for the project. This will demand effort, it will not happen as a matter of course, and it will require the dedication and motivation of people. The provision and training of an adequate management team is therefore an essential prerequisite for a successful job, for it is their drive and judgement, their ability to persuade and lead, which will ensure that the project objectives are achieved.

## 20.3  Project management – the way ahead

It is always difficult to prejudge the evolution and development of existing systems, particularly in the more subjective management area. The further into the future the target, the less can be seen, but prediction has to be attempted if progress is to be made. Prediction is based on an analysis of the existing situation and the related historic trends which are used as a mechanism to project into the future. This technique may be acceptable for many purely technological processes and for short-term predictions, but it is limited in its ability to encompass the wider picture.

As time goes on, new methods and techniques are developed and promoted with varying degrees of success. Some of these processes, for example total quality management and business process re-engineering, can be viewed in their own right, but in context can also be viewed as part of the overall process of project management. It is likely that further novel and innovative processes will be derived, and become incorporated into best practice for the delivery of world-class project management.

Project management relies upon good management practice, but has the overall goal of the project's completion in terms of the project objectives as its prime objective. The techniques mentioned above are also aimed at making improvements in management practice and hence increasing the effectiveness of projects, but they achieve this aim by concentrating on non-project parameters. The most successful users of these individual techniques are organisations not currently using project management for their internal and external projects, and often from a manufacturing or production background. Therefore, it is likely that project management will, over the next few years, subsume these non-project management approaches. This is likely to involve some changes in approach and terminology, but will result in improved project management procedures integrating discrete business functions to enhance the effectiveness of decision making.

Project management will be required until such time as all projects can be delivered to meet all the promoter's requirements, and any other project constraints, the first time and every time. Judging by current performance, there is likely to be a need for project management for the foreseeable future.

# Suggested Answers to Exercises in Chapter 9

## 9.1 New housing estate

The project consists of a small development of a new housing estate, familiar to many urban or suburban sites. The project includes four blocks of low-rise flats, a block of five-storey flats and a shops and maisonettes complex, with support services.

The key issue is to identify where most of the work is required. In this case the five-storey block absorbs the majority of the site man-hours, and unless careful assumptions are made regarding the gang sizes and the overlapping of the different stages of construction, it would not be possible to finish within the time allowed. Therefore, despite having blocks A and B as a priority, as soon as the drain has been diverted work must start on the five-storey block.

There is no single correct solution, as the precise answer depends upon the assumptions made. The model solution suggested is shown in Figure S1. Assumptions made include having two excavation teams of 8 men: the first team commences on block A and then goes to the five-storey block; the second team commences on block B and then also to the five-storey block. The concrete frame for the five-storey block is assumed to have 16 men, the brickwork 12 men and the finishers 20 men. It is further assumed that brickwork could overlap the frame by 14 weeks and that finishing could overlap brickwork by 15 weeks.

This results in a base demand for 12 bricklayers on site, rising to a plateau of 24 bricklayers for 20 weeks before falling back to 12 again. The logic is shown by dotted lines, and important constraints or key dates are clearly marked. The space within the bars has been used for figures of output. A histogram of the demand for bricklayers and for total labour has been plotted directly under the bars at the bottom edge of the programme.

This solution completes in 78 weeks, allowing 4 weeks float to compensate for an optimistic programme, and has a maximum number of

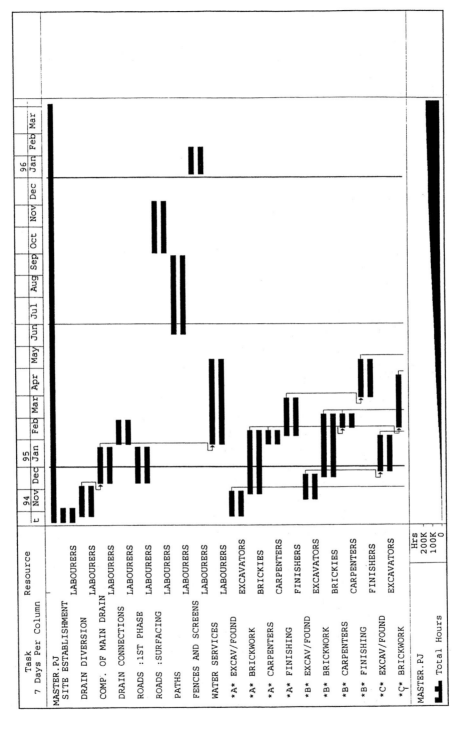

**Figure S1** New housing estate: suggested solution

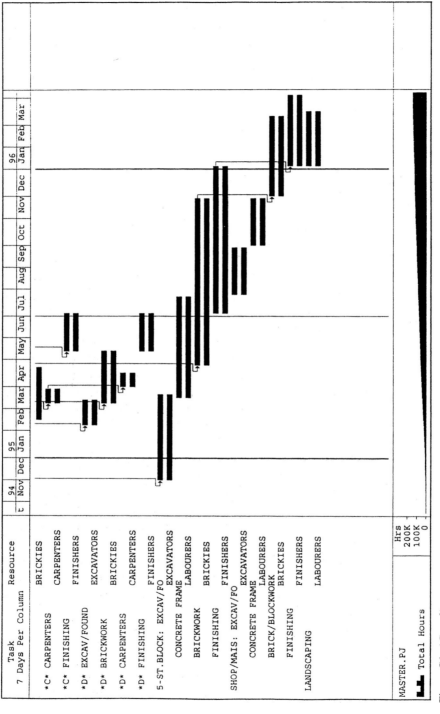

**Figure S1** *(Contd)*.

people on site of 65. Keeping the total on site low is important in practice, as accommodation and equipment has to be provided for all workers. There is a conflict between smoothing the bricklayers and a smooth demand for total labour, and computer-based methods are more efficient in resolving complex resources conflicts. With this solution flats A and B are ready by week 34.

It is important to note that for A and B to be habitable then assumptions have to be made about the proportions of work for the drain connections, paths, fences and screens, water services and landscaping that will have to be completed by week 34 also.

## 9.2 Pipeline

The time–location diagram is particularly suitable for cross-country jobs such as pipelaying, where the performance of individual activities will be greatly affected by their location and the various physical conditions encountered. In this exercise the assumption is that one stringing gang and four separate pipelaying gangs are to be employed. The remaining problem is to decide whether a single bridging gang can cope with all restrictions placed on the programme.

The programme has to adopt a trial and error approach to producing the diagram, to balance the needs for bridging with the productivity of the gangs. The time-location diagram, shown as Figure S2, shows a possible solution for one bridging gang, one stringing gang and four pipelaying gangs. Pipelaying gang 1 starts at 0 km, gang 2 at 9 km, gang 3 at 12 km and gang 4 at 16 km. It can be seen from the slopes of the progress lines, falling from 300 m/week/gang in normal ground to 75 m/week/gang in rock, that there is a need for two gangs to tackle the rock between 10 km and 13 km. The bridging gang move to keep ahead of each of the pipelaying gangs approaching an obstruction and hence move from the culvert at 2 km to the culvert at 18 km, to the thrust bore at 13 km, to the river crossing at 7 km, and finally to the railway crossing at 22 km.

Pipes are available at a maximum rate of 1000 m/week. Remembering the constraint to keep at least one week ahead of the pipelayers, a sequence for offloading from suppliers' lorries, storing and stringing out is shown on a weekly basis.

## 9.3 Industrial project

(a) The precedence diagram for this exercise is shown in Figure S3. The activities are represented by egg-shaped nodes and the interrelationships

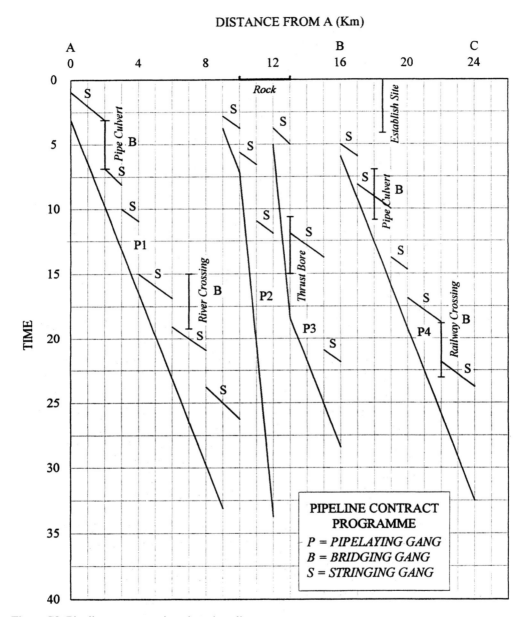

**Figure S2** Pipeline contract: time–location diagram.

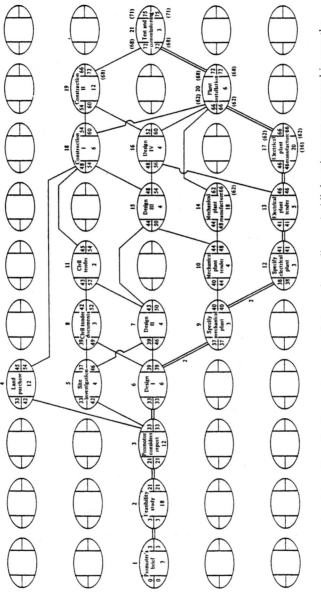

**Figure S3** Industrial project: Outline project network (precedence diagram). All durations are expressed in months.

between activities by lines known as *dependences*. The diagram is constructed and the forward and backward passes undertaken in exactly the same way as the worked example in the chapter. Minimum duration is 75 months and the critical path goes through activities 1–2–3–6–9–12–13–16–20–21.

*Note*: It is unlikely that you will get the most useful diagram at the first attempt. When redrawing a critical path diagram try to minimize the crossing of logical dependences, although often some crossing is unavoidable, but more importantly think about the role of the plan in communicating to the project team. In Figure S3 each level or row of the diagram represents a particular responsibility: top level, land purchase (legal department); second level, site investigation and civil (civil engineering); main level, project activities and design; fourth level, mechanical department; and fifth level, electrical department. This is one way of presenting the information, but there are many other ways that might be particularly useful in a given situation.

The float associated with activity 16 is as follows. *Total float*, the difference between its earliest and latest starts or finishes, is 8 months; *free float*, the minimum difference between the earliest finish time of that activity and the earliest start time of a succeeding activity, is 2 months.

(b) The option to spend more money to save time is frequently encountered in practical project management. In this simple example the combinations of A and C represent the existing situation and A and D, B and C, and B and D require investigation. However, activity 14 is not on the critical path, and therefore A and D would not be beneficial and can be ignored.

Take Activity 17 first and substitute B for A. Recalculate the remainder of the forward pass to show an early finish of 71 months. Calculating the backward pass provides two critical paths: 1–2–3–6–9–12–13–16–20–21 and 1–2–3–6–9–10–14–20–21. The extra cost of substituting B for A is £156 000 but the profit earned would be (75–71) 4 × £50 000 = £200 000, a gain of £44 000. Therefore option B should be adopted for activity 17.

Activity 14 is now on one of the critical paths. However, as there are two paths, a separate reduction in the duration of activity 14 would not reduce the project duration, and hence would not be cost effective; option C should be retained.

Therefore options B and C are the recommended choice.

## 9.4 Bridge

(a) Taking a step closer to the real world, no logic dependences are indicated, and the planner has to use judgement to construct the network. There are a number of possible networks, but it is recommended that the diagram should be kept as free and unconstrained as possible. It is suggested that 'set up site' would be the start activity, and this would then link to all the excavation activities. The piledriving for the right pier should be inserted after 'excavation' but before 'foundations'. All other 'excavation' activities are followed by 'foundations'. 'Foundations' are followed by the next stage of the 'Concrete' process. Next, beams can be placed, but remember in the logic that for a beam to be placed it must have the supports at both sides completed. Activity 18 'clear site' is suggested as the finish activity.

No figure is included, as there are a number of viable solutions. One of the optimal solutions gives completion in week 30 with two critical paths, both going through activity 6, the right pier pile driving.

(b) The effects of reducing the duration of pile driving to 5 weeks will vary depending upon the network drawn for part (a). Nevertheless, the overall effect on most networks is to reduce the completion time to 29 weeks with two different critical paths, one going through the north, and one the south, abutment activities.

(c) The resources are heavily constrained. One Excavation team, one concrete team for foundations, one concrete team for abutments and piers and one crane team severely limits the activities that can operate in parallel.

The question is slightly unfair, as this type of resource scheduling is difficult to undertake without the assistance of a computer. Many people find the use of resourced bar charts of assistance, but there is no easy way. In this case the obvious action is to employ resources away from the right pier such that the extra work required for the piledriving does not delay or disrupt the use of any resource.

If this is achieved, a completion time for a 'resource-constrained' project of 46 weeks can be achieved. However, any solution under 50 weeks represents a reasonable attempt at the exercise.

# Index

Note: Figures and Tables are indicated by *italic page numbers*